Food Science and Technology | Bulletin™

Functional Foods

Volume 4

ifis
INTERNATIONAL FOOD
INFORMATION SERVICE

Food Science and Technology Bulletin: Functional Foods comprises a minimum of eight minireviews per year. It is published as an annual review journal by IFIS Publishing, Lane End House, Shinfield Road, Shinfield, Reading RG2 9BB, UK. Tel. +44 118 988 3895. Fax +44 118 988 5065. Email ifis@ifis.org. Web: www.foodsciencecentral.com.

Charity registration no. 1068176.

ISSN 1476-2137
ISBN 978-0-86014-175-4

Purchase and subscription options
The *Food Science and Technology Bulletin: Functional Foods* is available to purchase as:

Individual minireviews
Minireviews may be purchased and downloaded individually from IFIS Publishing's website (www.foodsciencecentral.com/home/bulletin-ff) or from Atypon Link (www.atypon-link.com/IFIS/loi/fstb). An individual purchase gives unlimited access (by user name and password) to minireviews for one month. Individual minireviews cost US$44/£25/€35 each, plus VAT.

Annual subscriptions
An annual subscription provides unlimited access to all the published minireviews for one year, in addition to all previously published titles. Annual subscriptions for organisations cost US$309/£180/€263, plus VAT. Terms and conditions apply.
Annual subscriptions for personal use are sold with a 66% discount - US$104/£59/€86, plus VAT.

Food Science and Technology | Bulletin™
Functional Foods

Volume 4

Contents

Prebiotics in ancient diets

Jeff D. Leach

Paleobiotics Laboratory, 2401 Burgundy #7, New Orleans, LA 70117, USA

Abstract

While modern studies continue to expand our knowledge of the health benefits of prebiotics, virtually nothing is known of their use among ancient populations. Drawing on select ethnographical and archaeological data, examples of prebiotic use in ancient diet is presented. By utilising the well-documented facilities used to cook inulin-bearing plants that are found throughout the archaeological records of North America as a proxy, prebiotic consumption is documented in Europe and the Mediterranean possibly as early as 40 000 years ago. Data are further provided to suggest that early members of the genus *Homo* had ample ecological opportunity to include prebiotic underground storage organs found throughout the arid African savannah into the diet as early as 2.5 million years ago. This cursory view into the nutritional past of our ancestors reveals that prebiotics were likely to have been consumed in higher quantities than is the case among modern humans.

Keywords: archaeology, cook-stone technology, evolution, nutrition, prebiotics

1. Introduction

Since the 1970s, there has been a renewed interest in the link between colonic function and human health (Jenkins *et al.* 1999), with much recent attention being given to prebiotic carbohydrates that are not available for the vertebrate digestive system in general and for the human digestive system in particular, and as such are available in their entirety for the abundant intestinal bacterial ecosystem. Prebiotics interact in a selective way with the intestinal ecosystem and tend to change its composition with potential positive health effects for the consumer (Gibson and Roberfroid 1995; Gibson *et al.* 2004; Van Loo 2005). The well-established β(2-1) fructans inulin and oligofructose continue to drive much of the current research on the health benefits associated with prebiotics (Roberfroid 2002; Van Loo 2004a, 2004b). Although much current research is aimed at demonstrating health benefits for modern populations and the mechanisms for delivering prebiotics safely into the food supply (Franck 2002), very little is known about the consumption of inulin-type fructans throughout human history.

This paper briefly reviews archaeological evidence for prebiotic consumption in southern North America and select regions of the world. As a component of human health, it is useful to consider the evolutionary role of natural prebiotic foods from the perspective of nutritional ecology (Leach *et al.* 2006a, 2006b). This is defined as the study of essential nutrient intake for the purpose of overall human health, growth and maintenance, and general trends towards population growth (Jenike 2001; Jenkins *et al.* 1999). In other words, a diverse and sufficiently nutritional human diet will result in sustained or improved human health patterns, as revealed by lower infant mortality and an extension of human life expectancy.

The time-depth afforded by archaeology is unique in that it provides a window into dietary and other environmental variables that have shaped our current genetic make-up and its nutritional parameters. Within the last 10 000 years, significant changes nutritionally (i.e. due to agricultural changes) and technologically (i.e. as a result of the industrial revolution) have occurred too recently on a genetic time-scale for our genome to adjust (Cordain *et al.* 2002; Eaton *et al.* 2002; Goldsmith 1993; Williams and Nesse 1991). Thus, modern populations are selected biologically and physiologically for an evolution-based diet that did not include many of the popular foods that currently dominate diets. As such, the nature and composition of the modern gut microflora is in discordance and is progressively divergent from the original, genetically determined composition.

2. Evolution-based nutrition and nutritional ecology

Humans require a diverse diet of nearly 50 essential nutrients for proper growth, metabolic function and cellular

Food Science and Technology Bulletin: Functional Foods **4** (1) 1–8
DOI: 10.1616/1476-2137.14801. Published 25 May 2007

repair (IOM 2002). Current nutrient requirements and physiology have been conditioned by selective pressure and adaptability played out on an ever-changing nutritional landscape spanning millions of years. Fossil evidence places the earliest members of our genus (*Homo*) at approximately 2 million years ago (Finlayson 2005; Wood 2002). Throughout much of our history (>99%), humans evolved on a diet that was void of dairy foods, margarine (separated fats), cultivated cereal grains, and refined sugars, all of which supply as much as 60–70% of the calories in many modern diets. Up until approximately 500 generations ago, all humans consumed plants and animals foraged from their environment, and consumed virtually no agricultural grains or processed foods. Our evolution-based hunter-gatherer diet was high in fibre (dietary and functional), lean animal protein, polyunsaturated fats (omega-3 [ω-3] fatty acids), monounsaturated fats, vitamins, minerals, phytochemicals and antioxidants, and low in sodium (O'Keefe and Cordain 2004). Astonishingly, 'semi-modern' hunter-gatherers and less westernised groups that adhere more closely to this ancient diet and lifestyle than to more westernised diets are largely free from chronic degenerative diseases (Cordain *et al.* 2002; Shephard and Rode 1996) and biomarkers of illness such as rising blood pressure, increasing adiposity and insulin resistance (Blackburn and Poineas 1983; Glanville and Geerdink 1970; Joffe *et al.* 1971; Kuroshima *et al.* 1972; Merimee *et al.* 1972; Spielmann *et al.* 1982).

Though traditional hunter-gatherer diet and lifestyle vanished in its 'purest' form in the early 20th century (Murray *et al.* 2001), ongoing studies of diet and lifestyle among less-westernised groups that remain throughout the world are demonstrating that models of optimal nutrition (therapeutic diets) may be developed from these extant evolution-based diets. Within the medical community (Eaton 2007; Eaton *et al.* 2002), there is a slow but significant movement towards acknowledging that a conceptual framework for preventing diseases of affluence may be built upon a foundation constructed within evolutionary theory. At the core of this theoretical movement, often referred to as Darwinian medicine (Stearns 1999; Trevathan *et al.* 1999), is the idea that our current genetic pool was shaped by millions of years of natural selection in environments very different than the ones we live in today, and that much of our genetic make-up is based on a nutritional landscape that did not include the foods that currently dominate our westernised diet. The discordance between the rapid pace of our recent (last 10 000 years) cultural adaptations (agriculture, food processing technology) is far outstripping our biological (genetic) ability to keep pace.

While some single-gene mutations (e.g. against malaria) are examples of the speed at which natural selection can occur, the pathophysiology of many chronic diseases involves many more genes and much greater periods of time to evolve (Sing *et al.* 1996). While we are culturally and socially modern, driving around in hybrid cars for example, we are literally and biologically ancient hunter-gatherers.

Our modern requirements for a great number of essential nutrients to sustain health and well-being suggest this pattern developed early in our ancestral history. Humans, along with other extant hominoids (apes), evolved from a common plant-eating ancestor some 5 to 10 million years ago (Milton 1999). While orangutans, gorillas and chimpanzees have evolved on a diet of mainly fruits, leaves, flowers and bark, humans have developed a dietary path based on a mixture of plant and animal foods that has allowed for cerebral growth, gut anatomy and digestive kinetics. It is this diverse diet, and our ability to optimise it through intensification and technology, that makes us unique among all mammals.

Due to poor preservation of food remains in the archaeological record, it is difficult to derive exact macronutrient levels for food intake in a given diet for a specific region. However, field studies of the few remaining hunter-gatherer and foraging groups carried out during the early and mid-20th century provide some insight into the likely range and variability of our ancestral, evolution-based diet. In a comprehensive review of the ethnographical data on 229 hunter-gatherer and forager groups from all over the world, Cordain *et al.* (2000) suggested that the typical hunter-gatherer diet derived as much as 45–65% of total energy from animal food whenever and wherever possible, but that plant to animal ratios ranged from 35:65 to 65:35, depending on environment, season and latitude.

Clearly no single diet characterises the 'typical or best' hunter-gatherer and, by extension, ancestral, diet. Humans can, and do, thrive on a variety of diets. For example, the Australian aborigines are known to have eaten some 300 different species of fruit and 150 varieties of roots and tubers (Brand-Miller and Holt 1998; Gould 1980), while Alaskan Artic Eskimos are famous for a diet consisting almost exclusively of raw fat and protein from marine mammals (Ho *et al.* 1972).

In the 5–7 million years since bipedal primates appeared, nearly 20 species within the taxonomic tribe *hominin* have been identified in the fossil record, with only modern *Homo sapiens sapiens* still standing (Finlayson 2005; Wood 2002). At six billion strong, modern humans are clearly well-adapted and successful. Within nutritional ecology, the physical and biological success of our species, coupled with our genetically predetermined nutrient requirements and digestive physiology, indicate that a diverse diet of essential nutrients has characterised much of our history. As a cornerstone of modern health and nutrition, diverse diets are known to result in lower rates of infant mortality and an increased life expectancy (IOM

2002; Shuman 1996), both of which have a significant impact on population demographics.

Support for our diverse diet is found in the ethnographical and historical accounts among the 'relic' hunter-gatherer and foraging societies discussed above. The nutritional ecology approach suggests, due to their wide-spread occurrence among the world's flora and the direct evidence in archaeological records, that inulin-type fructans played an important role within a suite of essential nutrients in long-term health and the ultimate demographic success of our species.

3. Prebiotics in ancient diets

The discovery that the storage carbohydrate, fructan, occurs in a significant portion (>36 000 species) of the world's flora (Hendry 1987) all but guarantees that the now well-studied prebiotics, inulin and oligofructose, were consumed by our Pliocene and Pleistocene ancestors millions of years ago. As our early ancestors moved from the rainforest to the parched savanna woodlands of subtropical Africa, subsurface tubers, rhizomes, corms and perennial bulbs, many rich in prebiotics, would have been a ready and important source of energy (Hatley and Kappelman 1980; Laden and Wrangham 2005). Today, many of these same resources serve as staples for the modern foragers and farming groups that inhabit the same subtropical environs (Murray *et al.* 2001; Vincent 1985). However, digestion-inhibiting compounds and plant toxins present in many below-ground food sources would have limited their role as staples in the early diet of *Homo* until technological adaptations, such as fire, were introduced (Ragir 2000; Stahl 1984). Nevertheless, as early members of the genus *Homo* began their evolutionary march to mammalian dominance, the inclusion of prebiotics within a diverse and mixed diet would have no doubt conferred a selective advantage for the consuming population. As the archaeological evidence reveals, prebiotics have long been a part of the human diet and, in some areas and time periods, in quantities that far exceed those currently consumed by modern populations (Van Loo 1995).

The physical evidence for plant consumption by our early ancestors is virtually nonexistent, owing to poor preservation of organic plant parts in archaeological records; although stable isotope analysis of skeletal remains of early hominids is providing some insight into the quality and diversity of early diet (Lee-Thorp *et al.* 1994; Richards *et al.* 2001). For adequate preservation of prebiotic food evidence in the early human diet we must travel millions of years forward to the Upper Paleolithic (approximately 40 000 to 12 000 years ago) of Western Europe and the Mediterranean Basin, and to the early Holocene (approximately 10 000 years ago) of North America, before significant direct and indirect evidence of prebiotic food consumption becomes evident.

Decades of large-scale archaeological research in North America have documented extensive exploitation of prebiotic-rich plants such as agave (*Agave* spp.), sotol (*Dasylirion* spp), camas (e.g. *Camassia quamash*, *C. leichtlinii*) and wild onion (*Allium* spp.). While a great number of inulin-bearing plants were known as food sources among the prehistoric and historic groups of North America (Wandsnider 1997), these particular plants provide by far the oldest evidence of prebiotic consumption in North America, dating back over 9000 years (see Table 1).

In the Lower Pecos region of the Chihuahuan Desert in west Texas along the US-Mexican border, deeply stratified cave deposits document the uses of agave, sotol and onion that date back nearly 9500 years. Kept dry and preserved by the large overhangs that characterise many of the caves and shelters of the region, an extraordinary collection of human coprolites and preserved macro-botanical plant remains suggest that pit-baked prebiotic foods (e.g. agave, sotol, onion) were a mainstay of this desert economy (Sobolik 1990).

East of the Lower Pecos on the western edge of the Edwards Plateau in central Texas, the deeply buried Wilson-Leonard site has produced a 2 m diameter rock-lined earth oven that would have been used to cook the nutritious onion-like bulbs of camas (*Camassia* spp.). Charred camas bulbs recovered during excavation of the oven produced a date of approximately 8200 years ago (Bousman *et al.* 2002). However, although no charred camas bulbs were recovered from deeper excavations, 'stone-lined hearths' underlying the camas oven were dated to approximately 9410–9990 years ago, suggesting possible earlier use of prebiotics.

At the Stigewalt site in southeastern Kansas, remains of large (>2 m diameter), rock-filled earth ovens with charred onion (*Allium* spp.) bulbs were dated to approximately 8810–7910 years ago (Thies 1990). As with the large oven at the Wilson–Leonard site in central Texas, the occurrence of hand-excavated pits lined with pre-heated stones seem to be consistently associated with the cooking of prebiotic foods. This same pattern continues throughout the American Southwest, where thousands of agave-roasting pits (also known as *mescal* pits) are scattered about the landscape (Leach 2005). Similarly, in the American Northwest, large, rock-lined ovens were used to cook as much as 1500 kg of inulin-rich camas bulbs in a single firing event (Thoms 2003).

The reoccurring use of large, rock-lined earth ovens, which are often associated with cooking of inulin-rich plants (Wandsnider 1997), is well-documented in the historical and ethnographical records for North America and northern Mexico. For example, Castetter *et al.* (1938) describe cooking agave in pits among the Mescalero and Chiricahua Apache of the American Southwest:

Table 1. A sample of inulin-containing plants identified as food among indigenous populations (adapted from Wandsnider 1997)

Plant	Common name	Edible part	Reference
Liliaceae			
Allium ampeloprasum	Leek, great-headed garlic	Leaf, bulb	Incoll and Bonnett 1993
Allium cepa	Onion, shallot	Bulb	Darbyshire and Henry 1981
Allium porrum	Leek	Bulb	Darbyshire and Henry 1981
Allium sativum	Garlic	Bulb	Darbyshire and Henry 1981
Asparagus racemosus	-	Tuber	Incoll *et al.* 1989
Arthropodium milleflorum	-	Tuberous root	Incoll *et al.* 1989
Asphodelus aestivus	Asphodel	Tuberous root	Incoll *et al.* 1989
Bulbine bulbosa	-	Corm	Incoll *et al.* 1989
Caesia calliantha	-	Tuberous root	Incoll *et al.* 1989
Camassia spp.	Camas	Bulb	Yanovsky and Kingsbury 1938
Dichopogon strictus	-	-	Incoll *et al.* 1989
Erythronium gradiflorum	Glacier lily	Corm	Turner *et al.* 1990
Agavaceae			
Agave vera cruz	Agave*	Meristem	Meier and Reid 1982
Cordyline terminalis	Palm lily	Tuber	Meier and Reid 1982
Dracaena australis	-	Tuber	Meier and Reid 1982
Yucca filamentosa	-	Stem	Meier and Reid 1982
Asteraceae			
Arctium lappa	Edible burdock	Tap root	Incoll and Bonnett 1993
Cichorium intybus	Chicory	Tap root	Vukov *et al.* 1993
Cirsium oleraceum	Meadow cabbage	Tap root	Vukov *et al.* 1993
Helianthus tuberosus	Jerusalem artichoke	Tuber	Vukov *et al.* 1993
Inula helenium	Elecampane	Root	Incoll and Bonnett 1993
Microseris lanceolata	Murnong	Tuberous root	Incoll *et al.* 1989
Polymnia sonchifolia	Yacon	Tuber	Ohyama *et al.* 1990
Scorzonera hispanica	Black salsify	Tap root	Incoll and Bonnett 1993
Taraxacum officinale	Dandelion	Tap root	Yanovsky and Kingsbury 1938
Campanulaceae			
Campanula rapunculus	Rampion	Tap root	Incoll and Bonnett 1993
Boraginaceae			
Symphytum officinale	Comfrey	Leaf	Incoll and Bonnett 1993

*Over 300 species of agave have been reported throughout the American Southwest and northern Mexico (Gentry 1982), all of which are thought to contain inulin.

"Pits in which the crowns [*agave*] were baked were about ten to twelve feet in diameter and three or four feet deep, lined with large flat rocks... Upon this, oak and juniper wood was placed, and before the sun came up was set on fire. By noon the fire had died down, and on these hot stones was laid moist grass, such as bunch grass... The largest mescal crown was selected... they threw it in and threw the other crowns after it... After, the mescal [*agave*] had been covered with the long leaves of bear grass and the whole with earth to a depth sufficient to prevent steam from escaping."

In the American Southwest, ideal surface conditions and slow rates of soil accumulation, accompanied by repeated use of oven facilities and subsequent accumulation of oven debris (discarded cooking stones) over multiple seasons, has made it possible to map thousands of cooking facilities, which often reach over 1 m in height and cover areas tens of meters in diameter (Leach *et al.* 2005). The synthesis of hundreds of radiocarbon dates from cook-stone facilities across extensive areas of southern North America (Leach 2005) has revealed a steady increase in prebiotic food consumption beginning over 9000 years ago, culminating in substantial intensification around 1250 years ago. The intensification of prebiotic foods in southern North America (specifically the American Southwest) coincides with an increased reliance on cultivated crops such as corn (*Zea mays*), squash (*Cucurbita* sp.) and beans (*Phaseolus* spp.), and a large-scale growth in the human population. Therefore, while populations were making the transition to a diet heavily dependent on starchy cultivars, prebiotic foods played an important and often increasing regional role in this subsistence economy.

As we see in North America, the occurrence of cook-stone technology in the absence of recoverable plant remains may be used as a proxy indicator of the exploitation of prebiotic foods in the archaeological record. While a great number of foods are known to have been processed with a cook-stone, the occurrence of large (>1 m diameter), ovens are consistently associated with many prebiotic foods (Leach 2005; Wandsnider 1997).

Throughout Western Europe, similar remains of massive cooking facilities are known to occur in Wales, England, Scotland, Ireland and Scandinavia. Referred to locally as *fulacht fiadh*, recent urban development has led to the excavation of a number of these mounds, which can reach over a meter in height and several meters in diameter, representing dozens, if not hundreds, of individual oven events. While moist ground conditions have all but destroyed any evidence of the plants that may have been processed in these features, radiocarbon dates on small amounts of carbonised wood charcoal from the initial heating of cook-stones indicates that the majority of mounds were constructed within the last 6000 years. Similar cook-stone mounds of varying sizes, dating roughly within the same time period, are known in southern parts of Australia (Holdaway *et al.* 2002). As seen for North America, historical and ethnographical accounts of using large, hand-excavated pits and heated cook-stones is noted throughout Australia. In one example, between 1884 and 1850 British explorers observed the following among the people at Menindee on the Darling River:

"The oven is a hole dug into which are placed stones; a fire is then made and when the stones are become sufficiently hot, whatever fibrous things they eat, or animal, is put into this oven and covered over and a fire made over it, when it soon gets cooked (Brock 1988)."

Among the >800 plant foods known to have been eaten for tens of thousands of years by Aborigines in Australia (Brand-Miller and Holt 1998), many were tuberous roots and corms that contained prebiotic inulin (Van Loo 1995) and required prolonged cooking in rock-lined pits (Gott 1982; Gould 1980; Incoll *et al.* 1989).

By far the oldest known evidence of cook-stone technology (ovens) in Europe comes from the cave site of Abri Pataud in the Dordogne region of southern France. In excavations by a joint American-French team between 1958 and 1964, a series of cook-stone features, some >1 m in diameter, were dated to approximately 33 000–18 000 years ago (Movius 1963). While it is impossible to know if prebiotic plant tissue was processed in these ancient features, as no direct evidence in the form of plant material was reported, their use in cooking vegetable material is inferred from the overwhelming evidence of similar features recorded throughout the world.

In one final example, ancient cook-stone features were recently excavated on the southern Japanese island of Tanegashima (Dogome 2000). The oldest two features are buried 10 cm below a layer of Tane-4 volcanic ash, which is radiocarbon dated to about 30 500 years ago. One is a sandstone lens of about 0.75 m in diameter and the other is a sandstone-filled basin about 1.15 × 0.75 m in diameter that is underlain by carbon-stained sediment. Thermally-altered sandstone ranges in size from a few centimetres to a maximum of 25 cm. Similar cook-stone features and fire-cracked rock scatters were found in overlying deposits dated as late as 6500 years ago, including several features associated with 12 000-year-old Incipient Jomon pottery. Investigators concluded the Late Paleolithic cook-stone features and heavy stone tools were indicative of a plant-based diet (Dogome 2000). These cook-stone features, especially the basin-shaped forms, closely resemble remains of earth ovens found throughout western North America that were used to cook inulin-rich plant tissue (Leach 2005).

Although it is evident that our ancestors consumed large amounts of inulin-containing crops, it could be questioned whether the heat generated by means of cook-stone ovens or other cooking methods would destroy the inulin present in these plants. Direct tests in conditions mimicking cook-stone ovens have not been performed to date. In Louisiana and Northern Europe, inulin-containing chicory roots are roasted. The roots are spread on grids that are stacked in a particular building. Hot air that is generated by burning wood or coal is led through the roots, thereby heating them up to a temperature of 180°C (356°F). It was observed that under these conditions, between 10 and 20% inulin was degraded (Pazola and Cieslak 1979; Van Loo 1995). In cooking or frying experiments with inulin-containing food plants such as onions, it was shown that the losses of inulin were limited to 10% or less. From these observations it can reasonably be concluded that the heat treatment in the cook-stone ovens (<100°C, products not immersed in water) preserved the inulin content of the food plants very well, with expected losses of <10%.

4. Discussion

From the current discussion it is clear that our distant ancestors consumed, in varying quantities, plants containing prebiotic carbohydrates. These by definition are not digested in the upper intestinal tract, but rather interact in a specific way with the bacterial ecosystem that is abundantly present in the lower intestinal tract. Consumption of prebiotic carbohydrates such as inulin selectively promotes the growth of bacteria that are associated with a healthy condition (e.g. lactobacilli, bifidobacteria) and suppresses bacteria that are associated with disease (e.g. clostridia). At the same time the metabolic activity of the bacteria is stimulated, which results in the production of metabolites that are absorbed into the blood and exert beneficial effects in the rest of the body with, as a direct consequence: improved resistance to infection, better skeletal bone quality and a reduced risk of chronic diseases such as cancer, cardiovascular disease etc. (Van Loo 2004a, 2004b, 2005).

The interesting association between cook-stone technology and prebiotics offers some proxy of initial intensification, in the absence of the direct recovery of prebiotic plant tissue. Further, the durability of many of these cook-stone features makes their identification and possible utility in recognising large-scale patterns of prebiotic use across space and time feasible through inductive principles of investigation. We suspect that while our ancestors have always included amounts of prebiotic plants in their diet through daily foraging activities and some evidence for the use of cook-stones is present during the Middle Paleolithic (Mellars 1996), it was not until the onset of the Upper Paleolithic (approximately 40 000 years ago) with its ornaments, decorated tools, deliberate storage facilities, crudely tailored clothing, art and clear demographical pulses (Steiner 2002), that prebiotic plant foods began to play an increasing role in the dietary evolution of our species.

Increased demographic pressure resulted in shrinking territories, making access to preferred plants and high-return animal and aquatic resources less reliable. It is possibly as a result of this cultural pressure that the initial intensification (increased diet breadth) of under-utilised below-ground resources (tubers, bulbs), many rich in prebiotics, took place. This form of land-use intensification (Holly 2005; Thoms 2003) was the beginning of a long-term—albeit punctuated—prebiotic revolution, made possible by the adaptation of cook-stone technology. The evolutionary implications of prebiotic consumption on the development and relative success of our species is unknown, and requires further research. However, advances in processing technology brought about during the industrial revolution in the late 19th century, in conjunction with the increase in 'westernised diets' and accompanying medical maladies, have forever altered the delicate evolutionary-induced balance between food and human health, thereby resetting our metabolic and genetic clocks.

The concept of prebiotic food ingredients is an important development in nutritional research. Beyond local effects, the idea that prebiotics can selectively modulate gastrointestinal microbial fermentation to influence physiological processes that are known biomarkers of potential illness and human health is profound. However, in the case of even the best-designed human nutrition intervention trial, optimal controls may never be achieved, as the diet and lifestyle of—most likely all—members will differ significantly from their evolutionary-based and thus genetically determined optimal diet.

The future of prebiotic research may benefit from a better understanding of the essential nutrient profiles for humans that have evolved over millions of years of selective pressure and how these relate to intestinal health, as our evolutionary trajectory has arguably been towards maximising our adaptability—both physically and physiologically (Schlicting and Pigliucci 1998). In other words, the biological and physiological parameters of essential nutrients and their conditioning of human health are, for the most part, predetermined and grounded in our ancient past. The recent genome sequencing of *Bifidobacterium longum* (Schell *et al.* 2002) further points to a symbiotic and ancient relationship between our genus and the prebiotic plants on the landscape.

There is no doubt that the majority of the intermediate markers of disease risk and health that are currently being addressed using prebiotics and the modulation of intestinal flora have, at their source, multifactorial causes (Leach 2007). Evolution has the consequence that successful living organisms do best in those environments in which they were selected. As a consequence, an informed research agenda that includes an evolutionary perspective on 'ancestral' parameters of diet and microflora composition may advance the realisation and potential of future prebiotic research, with the aim being to achieve optimal health and nutrition. Through this research agenda, it may be possible to characterise the differences between modern and ancient intestinal health in relation to microflora composition in order to integrate microbiological, nutritional and epidemiological studies and data into an overarching paradigm that can serve to establish formulations resulting in effective recommendations for consumers.

5. References

Blackburn, H. and Poineas, R. 1983. Diet and hypertension: anthropology, epidemiology, and public health implications. *Progress in Biochemical Pharmacology* **19**: 31-79.

Bousman, C.B., Collins, M.B., Golberg, P., Stafford, T., Guy, J., Baker, B.W., Steele, D.G., Kay, M., Kerr, A., Fredlund, G., Dering, P., Holliday, V., Wilson, D., Gose, W., Dial, S., Takac, P., Balinsky, R., Masson, M. and Powell, J.F. 2002. The Paleoindian – archaic transition in North America: new evidence from Texas. *Antiquaries* **76**: 980-990.

Brand-Miller, J.C. and Holt, S.H.A. 1998. Australian Aboriginal plant foods: a consideration of their nutritional compositional and health implications. *Nutrition Research Reviews* **11**: 5-23.

Brock, D.G. 1988 [1844–6]. *To the desert with Sturt*, Royal Geographical Society of Australasia, Adelaide, Australia.

Castetter, E.F., Bell, W.H. and Grove, A.R. 1938. The early utilization and the distribution of Agave in the American Southwest. *University of New Mexico Bulletin, Biological Series* **5**(4). The University of New Mexico, Albuquerque, USA.

Cordain, L., Brand Miller, J., Boyd Eaton, S., Mann, N., Holt, S.H.A. and Speth, J.D. 2000. Plant-animal subsistence ratios and macronutrient estimations in worldwide hunter-gatherer diets. *American Journal of Clinical Nutrition* **71**. 682-92.

Cordain, L., Eaton, S.B., Miller, J.B., Mann, N. and Hill, K. 2002. The paradoxical nature of hunter-gatherer diets: meat-based, yet non-atherogenic. *European Journal of Clinical Nutrition* **56**: S42-S52.

Darbyshire, B. and Henry, R.J. 1981. Differences in fructan content and synthesis in some *Allium* species. *New Phytologist* **87**: 249-256.

Dogome, H. 2000. Summary (English). In: *The Yokomine C Site* (in Japanese): 1–2. Town Board of Education, Minami Tane, Kagoshima. Torai, Kagoshima, Japan.

Eaton, S.B. 2007. Preagricultural diets in evolutionary health promotion. In: Ungar, P.S., editor. *Evolution of the human diet: the known, the unknown and the unknowable:* 384-394. Oxford University Press, UK.

Eaton, S.B., Strassman, B.I., Nesse, R.M., Neel, J.V., Ewald, P.W., Willaims, G.C., Weder, A.B., Eaton II, S.B., Lindeberg, S., Konner, M.J., Mysterud, I. and Cordain, L. 2002. Evolutionary health promotion. *Preventive Medicine* **34**: 109-188.

Finlayson, C. 2005. Biogeography and evolution of the genus *Homo*. *Trends in Ecology and Evolution* **20**: 457-463.

Franck, A. 2002. Technological functionality of inulin and oligofructose. *British Journal of Nutrition* **87**: S287-S291.

Gentry, H.S. 1982. *Agaves of continental North America*. University of Arizona Press, Tucson, USA.

Gibson, G.R. and Roberfroid, M.B. 1995. Dietary modulation of the human colonic microbiota: introducing the concept of prebiotics. *Journal of Nutrition* **125**: 1401-1412.

Gibson, G.R., Probert, H.M., Van Loo, J.A.E., Rastall, R.A. and Roberfroid, M.B. 2004. Dietary modulation of the human colonic microbiota: Updating the concept of prebiotics. *Nutrition Research Reviews* **17**: 259-275.

Glanville, E.V. and Geerdink, R.A. 1970. Skinfold thickness, body measurements and age changes in Trio and Wajana Indians of Surinam. *American Journal of Physical Anthropology* **32**: 455-462.

Goldsmith, M.F. 1993. Ancestors may provide clinical answers, say 'Darwinian' medical evolutionists. *Journal of the American Medical Association* **269**: 1477-1480.

Gott, B. 1982. Ecology of root use by the aborigines of southern Australia. *Archaeology in Oceania* **17**: 59-67.

Gould, R.A. 1980. *Living archaeology*, Cambridge University Press, UK.

Hatley, T. and Kappelman, J. 1980. Bears, pigs, and Plio-Pleistocene Hominids: A case for the exploitation of belowground food resources. *Human Ecology* **8**: 371-387.

Hendry, G. 1987. The ecological significance of fructan in a contemporary flora. *New Phytology* **106**: 201-216.

Ho, K.-J., Mikkelson, B., Lewis, L.A., Feldman, S.A. and Taylor, C.B. 1972. Alaskan artic Eskimos: response to a customary high fat diet. *American Journal of Clinical Nutrition* **25**: 737-745.

Hockett, B. and Haws, J. 2003. Nutritional ecology and diachronic trends in Paleolithic diet and health. *Evolutionary Anthropology* **12**: 211-216.

Holdaway, S.J., Fanning, P.C., Jones, M., Shiner, J., Witter, D. and Nicholls, G. 2002. Variability in the chronology of late holocene aboriginal occupation on the arid margin of Southeastern Australia. *Journal of Archaeological Science* **29**: 351-363.

Holly, D.H., Jr. 2005. The place of "others" in hunter-gatherer intensification. *American Anthropology* **107**: 207-220.

Incoll, L.D. and Bonnett, G.D. 1993. The occurrence of fructan in food plants. In: Fuchs, A., editor. *Inulin and inulin-containing crops:* 309-322. Elsevier, Amsterdam, Netherlands.

Incoll, L.D., Bonnett, G.D. and Gott, B. 1989. Fructans in the underground storage organs of some Australian plants used for food by aborigines. *Journal of Plant Physiology* **134**: 196-202.

IOM (Institute of Medicine). 2002. *Dietary reference intakes of energy, carbohydrate, fiber, fat, fatty acids, cholesterol, protein, and amino acids*. National Academy Press, Washington, DC, USA.

Jenike, M.R. 2001. Nutritional ecology: diet, physical activity and body size. In: Panter-Brick, C., Layton, R.H. and Rowley-Conwy, P., editors. *Hunter-gatherers: an interdisciplinary perspective*: 205-238. Cambridge University Press, UK.

Jenkins, J.A., Kendall, C.W.C. and Vuksan, V. 1999. Inulin, oligofructose and intestinal function. *Journal of Nutrition* **129**: 1431S-1433S.

Joffe, B.I., Jackson, W.P.U., Thomas, M.E., Toyer, M.G., Keller, P. and Pimstone, B.L. 1971. Metabolic response to oral glucose in the Kalahari bushmen. *British Medical Journal* **4**: 206-208.

Kuroshima, A., Itoh, S., Azuma, T. and Agishi, Y. 1972. Glucose tolerance test in the Ainu. *International Journal of Biometeorology* **16**: 193-197.

Laden, G. and Wrangham, R. 2005. The rise of the hominids as an adapative shift in fallback foods: Plant underground storage organs (USOs) and australpith origins. *Journal of Human Evolution* **49**: 482-498.

Leach, J.D. 2005. Sharp increase in cook-stone use in the Chihuahuan Desert during periods of agricultural intensification. *Antiquaries* **79**. Available at: http://antiquity.ac.uk/ProjGall/leach05/.

Leach, J.D., Bousman, C.B. and Nickels, D. 2005. Comments on assigning a primary context to artifacts recovered from burned rock middens. *Journal of Field Archaeology* **30**: 201-203.

Leach, J.D., Gibson, G.R. and Van Loo, J. 2006a. Human evolution, nutritional ecology and prebiotics in ancient diet. *Bioscience and Microflora* **25**(1): 1-8.

Leach, J.D., Rastall, R.A. and Gibson, G.R. 2006b. Prebiotics: past, present and future. In: Gibson, G.R. and Rastall, R.A., editors. *Prebiotics: development and application:* 237-248. John Wiley and Sons Ltd., UK.

Leach, J.D. 2007. Evolutionary perspective on dietary intake of fibre and colorectal cancer. *European Journal of Clinical Nutrition* **61**: 140-142.

Lee-Thorp, J.A., van der Merwe, N.J. and Brain, C.K. 1994. Diet of *Australopithecus robustus* at Swartkrans from stable carbon isotope analysis. *Journal of Human Evolution* **27**: 361-372.

Lee, R.B. and Daly, R., editors. 1999. *The Cambridge encyclopedia of hunters and gatherers*. Cambridge University Press, UK.

Mellars, P. 1996. *The Neanderthal legacy: an Archeological perspective from Western Europe*. Princeton University Press, New Jersey, USA.

Meier, H. and Reid, J.S.G. 1982. Reserve polysaccharides other than starch in higher plants. In: Loewus, F.A. and Tanner, W., editors. *Encyclopedia of plant physiology, vol 13A*: 418-471. Springer, Berlin, Germany.

Merimee, T.J., Romoin, D.L. and Cavalli-Sforza, L.L. 1972. Metabolic studies in the African pygmy. *Journal of Clinical Investigation* **51**: 395-401.

Milton, K. 1999. A hypothesis to explain the role of meat-eating in human evolution. *Evolutionary Anthropology* **12**: 11-21.

Movius, H.L. 1963. The hearths of the Upper Périgordian and Aurignacian horizons at the Abri Pataud, Les Eyzies (Dordogne), and their possible significance. *American Anthropology* **68**: 296-325.

Murray, S.S., Schoeninger, M.J., Bunn, H.T., Pickering, T.R. and Marlett, J.A. 2001. Nutritional composition of some wild plant foods and honey used by the Hadza foragers of Tanzania. *Journal of Food Composition and Analysis* **13**: 1-11.

Ohyama, T., Ito, O., Yasuyoshi, S., Ikarashi, T., Minamisawa, K., Kubota, M., Tsukihashi, T. and Asami, T. 1990. Composition of storage carbohydrates in tubers of yacon (*Polymnia sonchifolia*). *Soil Science and Plant Nutrition* **36**: 167-171.

O'Keefe, J.H., Jr and Cordain, L. 2004. Cardiovascular disease resulting from a diet and lifestyle at odds with our Paleolithic genome: how to become a 21st-century hunter-gatherer. *Mayo Clinic proceedings* **79**: 101-108.

Pazola, Z. and Cieslak, J. 1979. Changes in carbohydrates during the production of coffee substitute extracts especially in the roasting process. *Food Chemistry* **4**: 41-47.

Ragir, S. 2000. Diet and food preparation: rethinking early hominid behavior. *Evolutionary Anthropology* **9**: 153-155.

Richards, M.P., Pettitt, P.B., Stiner, M.C. and Trinkaus, E. 2001. Stable isotope evidence for increasing diet breadth in the European mid-Upper Paleolithic. *Proceedings of the National Academy of Science* USA **98**: 6528-6532.

Roberfroid, M. 2002. Functional food concept and its application to prebiotics. *Digestion and Liver Disease* **34**: S105-S110.

Schell, M.A., Karmirantzou, M., Snel, B., Vilanova, D., Berger, B., Pessi, G., Zwahlen, M.C., Desiere, F., Bork, P., Delley, M., Pridmore, R.D. and Arigoni, F. 2002. The genome sequence of *Bifidobacterium longum* reflects its adaptation to the human gastrointestinal tract. *Proceedings of the National Academy of Science USA* **99**: 14422-14427.

Schlicting, C.D. and Pigliucci, M. 1998. Phenotypic evolution. A reactive norm perspective. Sinauer, Sunderland, Massachusetts, USA.

Shephard, R.J. and Rode, A. 1996. *The health consequences of modernization: evidence from circumpolar peoples*: 101-108. Cambridge University Press, UK.

Shuman, J.M. 1996. Nutrition in aging. In: Mahan, L.K. and Escott-Stump, S., editors. *Food, nutrition, and diet therapy*: 287-308. W.B. Saunders, Philadelphia, USA.

Sing, C.F., Haviland, M.B. and Reilly, S.L. 1996. Genetic architecture of common multifactorial diseases. In: Chadwick, D. and Cardew, G., editors. *Variation in the human genome*. Ciba Foundation Symposium **197**: 211-232. Wiley, Chichester, UK.

Sobolik, K.D. 1990. A nutritional analysis of diet as revealed in prehistoric human coprolites. *Texas Journal of Science* **42**: 23-36.

Spielmann, R.S., Fajans, S.S., Neel, J.V., Pek, S., Floyd, J.C. and Oliver, W.J. 1982. Glucose tolerance in two unacculturated Indian tribes of Brazil. *Diabetologia* **23**: 90-93.

Stahl, A.B. 1984. Hominid dietary selection before fire. *Current Anthropology* **25**: 151-157.

Stearns, S.C., editor. 1999. *Evolution in health and disease*. Oxford University Press, UK.

Steiner, M.C. 2002. Carnivory, coevolution, and the geographic spread of the genus *Homo*. *Journal of Archaeological Research* **10**: 1-63.

Thies, R.M. 1990. *The Archeology of the Stigewalt Site, 14LT351*. Kansas State Historical Society, Contract Archeology Series, Publication 7. Kansas State Historical Society, Lawrence.

Thoms, A.V. 2003. Cook-stone technology in North America: evolutionary changes in domestic fire structures during the Holocene. In: Frere-Sautot, M.-C., editor. *Colloque et experimentation: le feu domestique et ses structures au Neolithic aux Auges des Metaux*: 87-96 Collection Prehistories No. 9, Editions Monique, Mergoil, France.

Trevathan, W.R., Smith, E.O. and McKenna, J.J., editors. 1999. *Evolutionary medicine*. Oxford University Press, UK.

Turner, N.J., Thompson, L.C., Thompson, M.C. and York, A.Z. 1990. *Thompson ethnobotany: knowledge and usage of plants by the Thompson Indians of British Columbia*. Royal British Columbia Museum, Victoria, Canada.

Van Loo, J., Coussement, P., De Leenheer, L., Huybregs, H. and Smits, G. 1995. On the presence of inulin and oligofructose as natural ingredients in the Western diet. *Critical Reviews of Food Science and Nutrition* **35**: 525-552.

Van Loo, J. 2004a. Prebiotics promote good health. The basis, the potential and the emerging evidence. *Journal of Clinical Gastroenterology* **38**: S70-S75.

Van Loo, J. 2004b. The specificity of the interaction with intestinal bacterial fermentation by prebiotics determines their physiological efficacy. *Nutrition Research Reviews* **17**: 89-98.

Van Loo, J. 2005. Prebiotics: a nutritional concept gaining momentum in modern nutrition. *Food Science and Technology Bulletin: Functional Foods* **2**(8): 83-100.

Vincent, A.S. 1985. Plant foods in savanna environments: a preliminary report of tubers eaten by the Hadza of northern Tanzania. *World Archaeology* **17**: 131-148.

Vukov, K., Erdelyi, M. and Pichler-Magyar, E. 1993. Preparation of pure inulin and various inulin-containing products from Jerusalem artichoke for human consumption and for diagnostic use. In: Fuchs, A., editor. *Inulin and inulin-bearing crops*: 341-346. Elsevier, Amsterdam, Netherlands.

Wandsnider, L. 1997. The roasted and boiled: food composition and heat treatment with special emphasis on pit-hearth cooking. *Journal of Anthropological Archaeology* **16**: 1-48.

Williams, G.C. and Nesse, R.M. 1991. The dawn of Darwinian medicine. *Quarterly Review of Biology* **66**: 1-22.

Wood, B. 2002. Palaeoanthropology: hominid revelations from Chad. *Nature* **418**: 133-135.

Yanovsky, E. and Kingsbury, R.M. 1938. Analyses of some Indian food plants. *Journal of the Association of Official Agricultural Chemists* **21**: 648-665.

Synbiotics: making the most of probiotics and prebiotics by their combinations?

Delphine M.A. Saulnier

Department of Pathology, Baylor College of Medicine, Texas Children's Hospital, 6621 Fannin MC 1-2261, Houston, TX 77030, USA. E-mail dmsaulni@texaschildrenshospital.org

Abstract

A synbiotic is a combination of a live microbial food supplement (probiotic) and a prebiotic; specifically, the latter should be a substrate for the probiotic that acts to increase its survival/activity in the host gastrointestinal tract (GT) while also stimulating indigenous beneficial bacteria. This review gives a brief description of the attributes of a synbiotic and focuses on the advantages of adding a prebiotic in a probiotic product during its manufacturing process as well as the beneficial effects observed following synbiotic ingestion. During manufacture and storage, probiotic survival can be enhanced through the addition of a prebiotic. Evidence from simulated conditions of the GT and from animal models is accumulating to suggest that, following ingestion, synbiotics stimulate the growth of beneficial bacteria (i.e. bifidobacteria and lactobacilli) while improving the activity of probiotic strain(s). Recent human studies indicate that ingestion of synbiotics modulates the gut microbiota, promoting a healthier composition; it appears that synbiotics can be more efficient than either pro- or prebiotics alone in inducing this effect. Preliminary results have shown beneficial effects on biomarkers of diseases such as ulcerative colitis (UC) and colorectal cancers.

Keywords: synbiotic foods, probiotic bacteria, prebiotics, gastrointestinal tract, health foods

1. Introduction

1.1 Definitions

The protective nature of certain microorganisms, in particular lactic acid bacteria contained in foods and drinks, has a long history. To describe such microorganisms, the term probiotic was introduced in 1965 (Lilly and Stiwell 1965). Probiotics are currently defined as "live microorganisms which, when administered in adequate amounts, confer a health benefit on the host" (FAO/WHO 2002). Several studies have confirmed the efficacy of the probiotic approach in various disorders such as diarrhoea, allergies and irritable/inflammatory bowel diseases (Kolida *et al.* 2006). However, being ingested microorganisms, probiotics are not always able to survive host physicochemical protective barriers, such as stomach and bile acids. Once probiotics reach the lower gut, they have to compete for nutrients and an ecological niche with the established colonic microbiota. These factors can compromise probiotic establishment in the colon and therefore diminish efficacy.

Food Science and Technology Bulletin: Functional Foods **4** (2) 9–19
DOI: 10.1616/1476-2137.14833. Published 15 June 2007
ISSN 1476-2137

For this reason, an alternative approach—the prebiotic concept—has been introduced to target indigenous beneficial bacteria established in the host. Prebiotics are defined as "non-digestible food ingredients that beneficially affect the host by selectively stimulating the growth and/or activity of one, or a limited number of bacteria in the colon that can improve the host health" (Gibson and Roberfroid 1995). Targets for prebiotics are usually bifidobacteria and lactobacilli. However, the selectivity of prebiotics is not always fully established, and hence stimulation of beneficial genera alone may be difficult to achieve.

To alleviate the limitations of the probiotic and prebiotic approach, a solution could be to combine both in the form of a synbiotic. Synbiotics were defined around a decade ago by Gibson and Roberfroid (1995) as "mixtures of pro- and prebiotics, which beneficially affect the host, by improving the survival and implantation of live microbial dietary supplements in the gastrointestinal tract (GT)". The prebiotic should be a specific substrate for the probiotic, being able to stimulate its growth and/or activity while at the same time enhancing indigenous beneficial bacteria.

The term synbiotic refers indirectly to a synergy and that is why some authors have suggested that this term should refer exclusively to products in which the prebiotic compound selectively favours the probiotics (Schrezenmeir and de Vrese 2001).

1.2 Synbiotic market

Europe's probiotics market accounted for 1.4 billion euros, whilst the prebiotic sector was valued at 0.9 billions euros in 2005 (RTS 2005). Although there are no current figures for synbiotics, this market also seems to be growing rapidly (RTS 2005).

Within Europe, the dairy sector is the most developed segment of the probiotic market, with probiotic yoghurts and fermented milks being very popular. Consumer acceptance varies across Europe, with northern European and Scandinavian countries—those having a long history of traditional consumption of fermented dairy products—being the most developed markets. Probiotic dietary supplements have been slow to gain acceptance, but today new applications are constantly emerging. In the USA the situation is reversed, with dietary supplements by far the most accepted product format. Conversely, the USA dairy probiotics market is very underdeveloped.

Most synbiotics launched to date in the food industry have been dairy products. From a technical perspective, it is the survivability of the probiotic strains which dictates which synbiotics can be developed, as prebiotics can be used in most food applications. The environmental sensitivity of many probiotics currently limit their practical use in non-refrigerated foods (Crittenden *et al.* 2006). Powdered dietary supplements are another application of synbiotics (Searby 2006). Technical challenges are not the only barrier to development of the synbiotic market; consumer understanding presents an equally large obstacle. It seems that "consumers understand probiotics, but if you ask them about prebiotics, they do not understand them so well and they understand synbiotics even less" (Searby 2006). However, manufacturers have already included both probiotics and prebiotics in the same formulation, but may not market the product as a synbiotic.

Prebiotics are added to many foods, including yoghurts, cereals, breads, biscuits, milk desserts and ice-creams. The European prebiotic market is dominated by fructans and galactooligosaccharides as the active ingredients (RTS 2005). The probiotics included in most of the synbiotics sold in the European market belong to *Lactobacillus* and *Bifidobacterium* genera (Table 1). The probiotic concentration is not usually indicated. Prebiotics included in synbiotic products are mostly fructans (fructooligosaccharides (FOS) and inulin), with a concentration that varies between 0.5 and 2.0 g/100 g of product. It is very likely that some of the associations of prebiotics and probiotics have not been optimised; for instance, the fact that members of the *Lactobacillus* genus are able to ferment the recognised prebiotic inulin has only been reported for a few particular strains (Makras *et al.* 2005).

2. Beneficial effects associated with synbiotics

2.1 Improvement in the viability of probiotics during manufacture and storage through addition of prebiotics

Prior to ingestion by the consumer, probiotics are often added to a food product that is then stored. Technologies that can protect the viability of probiotics during manufacture and storage are thus highly desirable to enable the strains to be delivered in quantities sufficient to exert beneficial effects. The addition of a prebiotic may help to protect probiotics during manufacture and storage.

2.1.1 Protection by prebiotics during the manufacture of probiotic products

A few studies have investigated the effects of prebiotic addition during the manufacture of probiotic products in terms of an increase in the numbers/survival of the strain. The addition of 2% (w/v) FOS during the manufacture of yoghurt containing probiotic bifidobacteria (*Bifidobacterium lactis* Bb12 (Bb12) or *B. longum* Bb46) increased numbers

Table 1. Synbiotics in the European market (modified after Kolida *et al.* 2002)

Product	Company	Active ingredients
Actimel 0 % (yoghurt)	Danone (France)	*L. casei* Imunitass® + short-chain FOS
Activia (yoghurt)	Danone (France)	*B. digestivum*® + inulin
Aktifit (yoghurt)	Emmi (Switzerland)	LGG + FOS
Fyos (fermented drink)	Nutricia (Belgium)	*L. casei* + inulin
Fysiq (dairy drink)	Mona (Holland)	*L. acidophilus* + inulin
Probiotic plus Oligofructose (yoghurt)	Bauer (Germany)	*L. acidophilus* and *L. bifidus* type La7 + FOS
Simbi (yoghurt and drink)	Priegola (Spain)	*L. acidophilus* La5, *L. casei* LC01, Bb12 + FOS
Symbalance (yoghurt)	Tonilait (Switzerland)	*L. reuterii*, *L. acidophilus* and *L. casei* + inulin
Vitality (yoghurt)	Müller (UK)	*L. acidophilus* La5 and Bb12 + inulin

FOS: fructooligosaccharides; LGG: *Lactobacillus rhamnosus* GG; Bb12: *Bifidobacterium lactis* Bb12.

in the final product compared to a control without prebiotics (Akalin *et al.* 2004). This increase was probably due to partial fermentation of the FOS during manufacture. Instead of adding a commercial prebiotic, production of prebiotics (galactooligosaccharides) by different *Bifidobacterium* strains during a preincubation step in the production of yoghurt has been investigated (Lamoureux *et al.* 2002). Although production of galactooligosaccharides was modest (0.3–0.8% (w/v)), fermentation of the yoghurt culture and bifidobacteria that followed allowed the production of yoghurt in which bifidobacteria was present in relatively high cell numbers. Another study has shown that production of galactooligosaccharides during the manufacture of a plain culture yogurt (Dahi), containing mixed Dahi cultures and *Lactococcus lactis* ssp. *diacetylactis* NCDC60, maintained a sufficient number of probiotics *L. acidophilus* NCDC14 and *L. casei* NCDC19 in the final product (Yadav *et al.* 2007).

The manufacturing method of spray drying for delivering probiotics was investigated with the addition of prebiotics. During the manufacture process, the viability of *L. rhamnosus* GG (LGG) was monitored in the presence of different prebiotic substances (20% (w/v) inulin, oligofructose or polydextrose), which were included in the feed medium (Corcoran *et al.* 2004). However, presence of the prebiotics did not enhance the viability of the probiotic during spray drying when compared to reconstituted skimmed milk.

2.1.2 Protection by prebiotics during storage

Several factors have been claimed to affect the viability of probiotics in dairy products such as yoghurts, including acidity, pH, hydrogen peroxide, oxygen content, concentration of lactic and acetic acids and temperature of storage (Samona and Robinson 1994). A few studies have mentioned that the addition of prebiotics could improve the survival of the probiotic during storage. In general, in reconstituted skimmed milk containing either 5% (w/v) inulin, FOS, lactulose or resistant starch, fermented by *Lactobacillus* strains, survival of the probiotic following four weeks of storage at 4°C was optimal with inulin (Desai *et al.* 2004). Another study was carried out with *Bifidobacterium* strains, following the same protocol, and showed that in this case survival of probiotics was better in the presence of resistant starch (Bruno *et al.* 2002). These two studies highlight the fact that the choice of prebiotic(s) should depend on the probiotic used. When FOS, galactactooligosaccharides and inulin were compared, FOS was the most effective of the prebiotics tested in increasing the viability of bifidobacteria in skimmed milk during four weeks of storage at a comparable temperature (Shin *et al.* 2000). The effect was dose-dependent, being maximal at 5% (w/v) prebiotic. Addition of any of these prebiotics enhanced the survival of the strains compared to a control in the absence of prebiotic; activity of the strains, as measured by the production of acetic and lactic acids, was also higher.

In yoghurt, the addition of 2% (w/v) FOS also increased the viability of Bb12 and *B. longum* Bb46 following four weeks of storage at 4°C, compared to the respective control with no prebiotic addition (Akalin *et al.* 2004). The same conclusions were drawn from the study of Capela *et al.* (2006) in which the same dose of either FOS, inulin or resistant starch improved the viability of *L. acidophilus* 33200, *L. casei* 279, LGG and *B. longum* 536. It was also the case that addition of 0.25 or 2.5% (w/v) lactulose increased the viability of *B. bifidum* BB02 and *L. acidophilus* La-5 (Ozer *et al.* 2005).

Studies with synbiotic dairy products other than skimmed milk or yoghurts are sparse. Investigations have been performed to develop a synbiotic petit-suisse cheese, supplemented with *L. acidophilus* and *B. lactis* together with inulin, oligofructose and honey. Response surface methodology was applied to optimise probiotic viability, instrumental firmness, fructans content and cost (Cardarelli 2006). When a high dose (24% w/v) of inulin was used to supplement the petit-suisse cheeses, no increase in viability was seen for *L. paracasei* ssp. *paracasei* LBC 82 following three weeks of storage at 4°C when compared to a control without the prebiotic (Buriti *et al.* 2007). Another dairy product from the chilled food section—chocolate mousse—may also be a suitable vehicle for probiotics, despite the delicate processes involved in manufacture (Aragon-Alegro *et al.* 2006). Growth of yeasts and moulds may limit the shelf-life of this product: their growth was detected after 14 days for both the control and probiotic (*L. paracasei* LBC 82) mousses. Yeast growth for a synbiotic mousse containing *L. paracasei* LBC 82 and inulin was not observed until 21 days of storage at 5°C.

When considering the viability of spray-dried lactobacilli probiotics, no improvement was shown with the addition of prebiotic (inulin) after storage at 4°C compared to a control with no prebiotic added (Corcoran *et al.* 2004). The survival of an encapsulated *B. infantis* strain (in an oil-carbohydrate-protein emulsion) containing 20% (w/v) FOS, with or without 20% (w/v) resistant starch, and non-encapsulated *B. infantis* were monitored during storage at 25°C over a five week period, which showed that encapsulation improved survival (Crittenden *et al.* 2006). However, the mechanisms leading to such a protection were not explained and a negative control capsule containing FOS was not included in the study. As such, the role of the prebiotics in mediating this protection cannot be ascertained.

There is some evidence that the addition of prebiotics can improve the survival of probiotics during the storage of probiotic-containing foods, especially fermented milk products, but the mechanisms of effect are unclear. Hypotheses can be made that beneficial results obtained may be due to a partial fermentation of the carbohydrate, as an increase of lactic and acetic acids during storage are sometimes reported. Results may depend upon the choice of the combination of pro- and prebiotics, their dose, as well as the food matrix into which they are incorporated.

2.2 Beneficial effects of synbiotics in vitro

2.2.1 Protection by prebiotics during simulated conditions of the GT

To be effective in the colon, probiotics should remain viable in the food product and also survive during passage through the GT. The acidic environments of the stomach and inhibitory effects of bile acids secreted into the duodenum are possibly the major impediments to the survival of ingested bacteria during intestinal transit. Lactobacilli and bifidobacteria have been shown to be sensitive to acidic environments and bile acids (Liong and Shah 2005; Vernazza *et al.* 2006).

In vitro, it has been reported that the presence of different prebiotics could enhance the survival of probiotics. Addition of 5 g/L FOS alleviated the toxic effect of bile acid when added to pure cultures of three *Bifidobacterium* strains (Perrin *et al.* 2000). Soygerm powder (4 g/L), containing the potential prebiotic soyoligosaccharides, as well as other compounds, increased the resistance of *L. reuteri* to bile salts and decreased membrane damage (De Boever *et al.* 2001). The presence of resistant starch increased survival of *Bifidobacterium* strains, which were able not only to use this substrate, but also to adhere to the starch granules (Wang *et al.* 1999), suggesting that bacteria attached to particles are more tolerant to hostile environments. Bifidobacteria encapsulated in a coating containing 20% (w/v) FOS offered better protection in a two-stage *in vitro* model simulating the conditions of the human stomach and small intestine, compared to a control without any encapsulation (Crittenden *et al.* 2006). However, the mechanism for this protection cannot be attributed with certainty to the prebiotic as the capsule also contained other materials.

2.2.2 Modulation of the gut microbiota by synbiotics *in vitro*

A few studies have investigated the effects of synbiotics on the gut microbiota *in vitro*, in batch cultures inoculated with faeces or models simulating the human gut (Gmeiner *et al.* 2000; Fooks and Gibson 2003; Pereira *et al.* 2003; Tzortzis *et al.* 2004). In batch culture, a comparison was performed between a prebiotic alone and a synbiotic, but there are generally a lack of comparisons with probiotics alone. Survival of the probiotic has often been determined at the genus and not at the strain level in batch culture (Fooks and Gibson 2003; Tzortzis *et al.* 2004). In the study of Fooks and Gibson (2003), higher numbers of lactobacilli were seen in a synbiotic combination containing FOS and *L. plantarum* 0407, compared to the prebiotic alone. The synbiotics tested inhibited the growth of pathogenic *Campylobacter jejuni* and *Escherichia coli* better than the prebiotic alone. In the study of Tzortis *et al.* (2004), a higher increase in lactobacilli and bifidobacteria was seen when added with a potential prebiotic (galactosyl melibiose mixture) compared to the probiotics alone. An increase in lactate and acetate in the synbiotic compared to the prebiotic was also attributed to the presence of the probiotic and its activity. This is one of few studies with a synbiotic tested in batch culture whereby a proportion of the microbiota was enumerated using molecular techniques (fluorescent *in situ* hybridisation (FISH)).

In a three-stage continuous culture system representing the human colon, experiments with a synbiotic carried out by Pereira *et al.* (2003) showed no increase in the probiotic *L. fermentum* KC5b in the presence of galactooligosaccharides, even though this strain was able to ferment the substrate in a pure culture. In the study by Gmeiner *et al.* (2000), a synbiotic composed of *L. acidophilus* 74-2 in combination with FOS increased the level of bifidobacteria in a simulation of the human intestinal microbial ecosystem. Survival of the probiotic was only estimated at the genus level and no comparison was made with either the probiotic or prebiotic in isolation.

Little evidence has been provided for a role for synbiotics in modulating the gut microbiota *in vitro*. In general, effects were similar to those with the prebiotic, especially increases in bifidobacteria and/or lactobacilli. The probiotic strain sometimes increased in numbers because of the presence of the prebiotic, but not always. In these studies it is difficult to attribute an effect of the synbiotic to each of its compounds as there is generally no comparison performed with the probiotic alone.

2.3 Beneficial effects of synbiotics in animals

2.3.1 Improvement in survival of probiotics in the GT

One desirable attribute of a synbiotic is to improve survival of the probiotic through the GT, and some studies have determined this aspect in animals.

A six-fold greater recovery of the strain *Bifidobacterium* spp. Lafti$^{(TM)}$ 8B was noted in mouse faeces following an oral dosage when resistant starch was added for two weeks, compared to a control without this substrate (Wang *et al.* 1999). This strain was previously shown to be able to ferment this substrate and adhere to the granules of resistant starch *in vitro*. These two characteristics could have protected the strain during transit through the GT of the mice, and thus enhanced its survival. Increases in numbers of lactobacilli were seen following administration of a *Lactobacillus* strain along with 3 g/day FOS for 10 days in piglets, compared to controls fed the probiotic alone. However, in this case recovery was not monitored at the strain but at the genus level and it is possible that addition of the prebiotic had increased species of lactobacilli other than the probiotic itself. An increase in survival of the probiotic was not always seen; for example, no improved recovery of *B. breve* administered to mice with 10 g/day *trans*-galactooligosaccharides was recorded in one study (Asahara *et al.* 2001).

2.3.2 Modulation of the gut microbiota

In rats, administration of *Bifidobacterium* strains (*B. longum* KN29.1, *B. longum* KNA1, *B. animalis* KSp4) alone or in combination with FOS (5% (w/v) of the diet) showed a bifidogenic effect of the prebiotic and synbiotic combinations (Bielecka *et al.* 2002). Survival of the strains was not monitored, but an increase in bifidobacteria was no higher with the synbiotic compared to the prebiotic alone.

2.3.3 Protection against pathogens

An improved protection against pathogens by a synbiotic, compared to prebiotic and prebiotic administered alone, has been shown in a *Salmonella enterica* serovar *Typhimurium*-induced infection in mice (Asahara *et al.* 2001). In this study, *B. breve* was administered with *trans*-galactooligosaccharides for seven days before challenge with *Salmonella*, and for another seven days afterwards. An increased production of short-chain fatty acids (SCFA) and lower pH in the intestine was correlated with an inhibitory effect on the pathogen and may be due to fermentation of the prebiotic by the probiotic strain. Although the probiotic was recovered in the same quantity with and without the prebiotic, it appears that its activity was enhanced by fermentation of the prebiotic.

2.3.4 Decreased risk of colon cancer

Interest in the decreased risk of colon cancer has been generated with synbiotics and results seem encouraging in animals. Aberrant crypt foci (ACF), used as an index of colon cancer, were monitored following ingestion of *B. longum*, inulin or a combination of both in rats given the carcinogen azoxymethane (AOM; Rowland *et al.* 1998). The synbiotic resulted in a greater inhibition of small ACF (80%) compared to ingestion of the probiotic (26%) or prebiotic (41%) alone. A similar experiment was performed studying the effects of ingestion of either a *Bifidobacterium* strain, FOS (prebiotic) or both in rats given the carcinogen 1,2 dimethylhydrazine (Gallaher and Khil 1999). Neither the probiotic nor the prebiotic alone led to reduced numbers of ACF, but a reduction in ACF were seen following ingestion of the synbiotic. In long-term experiments, AOM-treated rats were fed either probiotics (LGG and Bb12), prebiotic (FOS-enriched inulin), or a combination of both for 31 weeks (Femia *et al.* 2002). Results showed that the synbiotic group had fewer tumour-bearing rats than the probiotic or prebiotic groups. A higher concentration of SCFA and molar ratio of butyrate was seen in the prebiotic and synbiotic groups. The same protocol used in the study of Femia *et al.* (2002) with AOM-treated rats was applied in the study of Roller *et al.* (2004a), and results showed that suppression of the immune system, due to AOM treatment, was counterbalanced following ingestion of prebiotic or synbiotic, but not probiotic substances.

In general, synbiotics were the preferred combination for a reduction in markers of risk for colon cancer in AOM-treated rats. However, in nearly all of the studies reported to date the same synbiotic was tested. It would therefore be interesting to measure the efficiency of other combinations of prebiotic and probiotics.

2.3.5 Inflammatory bowel diseases

Synbiotics have been shown to improve colitis in different animal models, such as dextran sulfate sodium-induced colitis rats (Osman *et al.* 2006) and HLA-B27 transgenic rats (Schultz *et al.* 2004). In the former study, a combination of *B. infantis* (DSM 15158 or DSM 15159) with oligofructose-enriched inulin improved the disease index and had anti-inflammatory effects. However, the effects of the synbiotic were different depending on the probiotic used. In the latter study, a combination of *L. acidophilus* La5 and Bb12 with inulin resulted in a measurable impact on the caecal microflora profiles of transgenic rats, by increasing the diversity and stimulating specifically the growth of *B. animalis*. Moreover, the colitis was attenuated.

2.3.6 Stimulation of the immune system

Other effects investigated with synbiotics in animals include stimulation of the immune system in healthy rats (Roller *et al.* 2004b). Synbiotics (Bb12, LGG and oligofructose-enriched inulin) increased secretory immunoglobulin A production in the ileum and natural killer cell activity in peripheral blood. It was concluded that effects of the synbiotic on the immune system were not simply a result of the additive effect of the prebiotic and probiotic.

2.3.7 Other effects

Investigation of the digestibility of B-vitamins when given along with probiotic, prebiotic or synbiotic to pigs with ileo-rectal anastomosis for 12 days has been carried out, but no effect was observed with any of these products (Branner and Roth-Maier 2006).

In rats a combination of LGG and Bb12 together with inulin has been shown to have an effect on two plasmatic neuropeptides, which inhibit both gastric emptying and acidic secretions (Lesniewska *et al.* 2006). Interestingly, in this study the synbiotic had opposite effects in adult and elderly rats, which suggests that dietary modifications with synbiotics depend on the age of the animals studied.

Animal studies with synbiotics have shown that survival of a probiotic in the GT can be improved through addition of a prebiotic, but this was not seen in all studies investigating this effect. The main beneficial effects of synbiotics recorded were an increase in bifidobacteria/lactobacilli and a decrease in pathogens. There is strong evidence in animal models to support the observation that synbiotics can decrease the risk of colon cancer, more efficiently than through ingestion of prebiotics or probiotics alone.

2.4 *Effects of synbiotics in humans*

2.4.1 Case studies

Case studies were among the first reports of the effects of synbiotics in humans. Synbiotic therapy has mostly been used to establish a 'healthy' microbiota, demonstrated in studies of between one and seven patients suffering from different infections in the GT (Kanamori *et al.* 2001; Kanamori *et al.* 2002; Kanamori *et al.* 2003; Kanamori *et al.* 2004) or with short bowel syndrome (Uchida *et al.* 2007). The same synbiotic was used in all of these studies, composed of *L. casei* (strain Shirota) and *B. breve* at a dose of 10^9cfu/day, and GOS at a dose of 3 g/day. Synbiotics were administered for at least nine months. Recovery of the probiotics was monitored in faecal samples following ingestion and showed that strains survived through the GT. In general the synbiotics were well tolerated and an improvement in the intestinal flora was noted, with a decrease in pathogenic species and an increase in lactobacilli and bifidobacteria, as enumerated by culture methods. All of these case studies lacked a negative control because of the low number of patients included. However, this synbiotic therapy seemed promising as the patients had fewer infections as a result and, moreover, the composition of their gut flora improved.

2.4.2 Human trials

The concept of synbiotics is recent and this explains why most of the intervention studies in humans have been carried out during the last 3–4 years. Human trials with synbiotics are presented in Table 2.

Some trials in humans have been carried out for the evaluation of a 'synbiotic', but are not reported here as the synbiotic did not fulfil the formal definition; in particular the carbohydrates were not recognised as prebiotic, as in the case of dextran (Ogawa *et al.* 2006) or lactose (Passeron *et al.* 2006). Other compounds, such as vitamins and minerals, were also included with the synbiotic (Bunout *et al.* 2002; Morelli *et al.* 2003) making results on the effect of the synbiotic inconclusive. Combinations tested in the synbiotics included mostly *Lactobacillus* and *Bifidobacterium* commercial strains for the probiotics, such *as* LGG and Bb12, and the already well-recognised prebiotics, such as fructans or galactooligosaccharides. Very often a combination of several probiotics has been used, and more seldomly a combination of prebiotics. The choice of combinations of prebiotic and probiotics was not justified in some cases as it was reported only in a few publications that the probiotics chosen were able to ferment the prebiotic added. Doses of probiotics were similar in the study, ranging from 4×10^9 to 10^{10} CFU/day. Doses of prebiotics ingested were between 8 and 20 g/day.

Populations targeted by these trials are diverse and include healthy subjects, but were carried out mostly with patients suffering from different clinical conditions (e.g. with diseases such as UC or colon cancer, or those that were polypectomised or critically ill, or with minimal encephalopathy cirrhosis or before liver transplantation). A pharmaceutical synbiotic (Synbiotic2000 or Synbiotic 2000Forte, Medipharm, Sweden) composed of four probiotics strains and four prebiotics has been tested fairly successfully in a low number of trials with critically ill patients (Koztampassi *et al.* 2006; Rayes *et al.* 2005; Voudouris *et al.* 2005), but had no effect in patients with Crohn's disease (Chermesh *et al.* 2007).

Survival of the probiotics were rarely measured in the presence of the prebiotic in the different trials, except in the study of Alander *et al.* (2001) where survival of the probiotic Bb12 was monitored through random amplification of a polymorphic DNA technique and showed an equal survival of the strain with or without the prebiotic. In another trial, survival of the probiotics with the prebiotic was investigated in a small number of volunteers prior to the trial using rifampicin-resistant variant probiotics (Van Loo *et al.* 2005). These studies excepted, most of the trials in humans are inconclusive in terms of whether or not the presence of the prebiotic increased survival of the probiotic in the GT.

Regarding the persistence of probiotic strains in the colon after the end of a feeding trial due to the presence of prebiotic, the study of Bouhnik *et al.* (1996) suggested that persistence of the *Bifidobacterium* strain was better in a synbiotic group, as the number of bifidobacteria were higher two weeks following the end of the treatment compared to the group that had ingested prebiotic alone. In contrast, in the study of Alander *et al.* (2001), persistence of Bb12 was low and found to be identical in the probiotic

Table 2. Summary of human studies with synbiotics

Probiotic(s)	Prebiotic(s)	Model	Duration	Effects	Comp	Authors
Bifidobacterium spp.	Inulin* 18 g/d	12 healthy volunteers, P-controlled	12 days	Inulin did not enhance increase Bif with SYN. Higher number of Bif. 2 weeks post-treatment with SYN.	PRO P	(Bouhnik *et al.* 1996)
Bb12 3×10^{10} CFU/d	GOS* 8 g/d	30 healthy volunteers, randomised	2 weeks	Increase of lactic acid bacteria with all the combinations. Increase of Bb12 with PRO and SYN. GOS did not enhance survival/persistence of Bb12. Bb12 may transiently replaced *B. longum* species, as measured by PCR-ELISA.	PRE PRO	(Malinen *et al.* 2002)
Bb12, La5, 4×10^{9} CFU/d	FOS 15 g/d	90 critically ill patients double-blind P-controlled	10 days	Decrease of carriage of pathogenic bacteria.	P	(Jain *et al.* 2004)
Bb12, La5, 4×10^{9} CFU/d	FOS 16 g/d	144 critically ill patients, P-controlled	variable	No measurable effects on bacterial translocation, gastric colonisation, systemic inflammation, septic complications.	P	(Anderson *et al.* 2004)
4 probiotics ◆ 10^{10} CFU/d	4 fibres ◇* 20 g/d	45 patients with cirrhosis, randomised P-controlled	30 days	Decrease of *E. coli* staphylococci with SYN and PRE. Increase of Lab with SYN, of Bif with PRE.	PRE P	(Liu *et al.* 2004)
4 probiotics ◆ 10^{10} CFU/d	4 fibres ◇* 20 g/d	66 patients scheduled for liver transplantation	2 weeks	Decrease of duration of antibiotic therapy and incidence of nosocomial infection with SYN compared to PRE.	PRE	(Rayes *et al.* 2005)
B. longum 10^{10} CFU/d	FOS/inulin* 12g/d	16 patients with active ulcerative colitis, double-blind randomised P-controlled	1 month	Increase of Bif. in the mucosa by quantitative PCR and reduction of inflammatory markers and colitis	P	(Furrie *et al.* 2005)
B. bifidum Bb02 *B. lactis* Bl01 3.5×10^{10} CFU/d	FOS/inulin 6g/d	18 elderly, double-blind randomised P-controlled	4 weeks	Increase of *B. bifidum*, *B. lactis* and Bif. with SYN, as measured by quantitative PCR. Detectable level of probiotics species 3 weeks post-treatment with SYN.	P	(Bartosch *et al.* 2005)
4 probiotics ◆ 10^{11} CFU/d	4 fibres ◇* 10 g/d	critically ill	2 weeks	Favourably altered the systemic inflammatory response and seemed to be associated with measurable clinical benefits.	PRE	(Voudouris *et al.* 2005)
4 probiotics ◆ 10^{11} CFU/d	4 fibres ◇* 10 g/d	65 critically ill trauma patients, randomised, P-controlled	variable	Reduced rate of infections, systemic inflammatory response syndrome, severe sepsis and mortality with SYN.	P	(Kotzampassi *et al.* 2006)
L. casei Shirota 4×10^{10}, *B. breve* 1×10^{10}	GOS, 15 g/d	81 patients undergoing high-risk hepatobiliary resection, randomised	2 weeks	Preoperative oral administration of SYN can enhance immune responses, attenuate systemic postoperative inflammatory responses and improve intestinal microbial environment.		(Sugawara *et al.* 2006)
S. boulardiii 2.5×10^{9} or 5.0×10^{9} CFU/d	Lactulose10 or 15g/d	30 healthy adults, cross-over randomised double-blind P-controlled	4 weeks	Bif increased with PRE and SYN as measured by denaturing gel gradient electrophoresis and quantitative PCR.	PRE PRO SYN	(Vanhoutte *et al.* 2006)
4 probiotics ◆ 10^{10} CFU/d	4 fibres ◇* 10 g/d	30 patients with Crohn's disease, randomised double-blind P-controlled	2 years	No effect on postoperative recurrence of patients with Crohn's disease.	P	(Chermesh *et al.* 2007)
Bb12, LGG 10^{10} CFU/d	FOS/inulin 10 g/d	80 patients (43 polypectomised, 37 with colon cancer), randomised double-blind P-controlled	12 weeks	Several colorectal cancer biomarkers were altered favourably by SYN. Bif increased and coliforms decreased with SYN. Supplementation with this SYN has minor stimulatory effects on the systemic immune system.	P	(Rafter *et al.* 2007; Roller *et al.* 2007; Van Loo *et al.* 2005)

Bif: bifidobacteria; Bb12: *B. lactis* Bb12; Comp: comparisons; ELISA: enzyme-linked immunosorbent assay; FOS: fructooligosaccharides; GOS: galactooligosaccharides; Lab: lactobacilli; LGG: *L. rhamnosus* GG; P: placebo; PCR: polymerase chain reaction; PRE: prebiotic; PRO: probiotic; SYN: synbiotic; *: was shown to be fermented by the probiotic(s) ; ◆: *P. pentosaceus* + *L. mesenteroides* + *L. paracasei* + *L. plantarum* ; ◇: inulin + β-glucans + pectin + resistant starch.

and synbiotic groups, with 20% of the volunteers still harbouring the strain two weeks following the end of the trial. Results may depend on the probiotic/prebiotic used in these studies.

Modulation of the gut microbiota was the most common parameter measured in these studies. In general the results, mainly obtained from culture methods, showed an increase in the number of bifidobacteria during synbiotic ingestion. A decrease in pathogenic bacteria or a reduction in the occurrence of infections was observed in studies with patients suffering from infections (Jain *et al.* 2004; Rayes *et al.* 2005). In nearly all of the studies, analysis of a small part of the microbiota was provided; a few groups/species, generally not dominant, were enumerated by plate counting.

Synbiotics were proven successful in reducing inflammatory markers and colitis in active UC patients in a pilot study (Furrie *et al.* 2005). Although only 18 volunteers took part in this study, the promising results obtained should lead to larger trials being conducted with the same synbiotic combination in patients suffering from inflammatory bowel disease (http://clinicaltrials.gov/ct/show/NCT00305409?order=1).

There is also hope that synbiotic therapy could be beneficial in colon cancer patients. In particular, this has been evaluated within the EU-funded project SYNCAN (www.syncan.be) in which a synbiotic was developed and ultimately tested in a large, randomised, double-blind, placebo-controlled, cross-over study with 80 polypectomised or colon cancer patients. The synbiotic, composed of the probiotic LGG and Bb12, together with oligofructose-enriched inulin, favourably altered biomarkers of colon cancer and increased bifidobacteria but had only minor stimulatory effects on the systemic immune system (Van Loo *et al.* 2005; Rafter *et al.* 2007; Roller *et al.* 2007).

Synbiotics targeted at elderly and healthy adults have been evaluated in large, randomised, double-blind, placebo-controlled, cross-over studies in other EU projects such as CROWNALIFE (www.crownalife.be) or EUMICROFUNCTION (www.eumicrofunction.be). Preliminary results from both trials showed an increase in bifidobacteria and beneficial effects upon selected markers of health (www.crownalife.be/html/latestresultsframeset.html and www.eumicrofunction.be/site/latest_frameset.html; Saulnier 2006)

Comparisons between studies performed so far is difficult as the probiotic/prebiotic and their respective doses, as well as the duration of the study, the population targeted and effects expected, are varied. In general, well-designed, randomised, double-blind, placebo-controlled studies are needed to link the observed beneficial effects in relation to the composition of the gut microbiota/survival of the probiotic(s) on one hand and to determine the dose of pro- and prebiotics necessary to obtain these effects on the other.

3. Conclusions

Promising results obtained during the manufacture of a synbiotic product in terms of increased probiotic survival, as well as after ingestion of synbiotics and from *in vitro* and both animal and human studies, warrant further investigations within the next few years. Degradation of prebiotics by the probiotic(s) should be a prerequisite when developing a synbiotic, in order to exert good synergy between the two compounds and maximise beneficial effects. A better understanding of prebiotic metabolism could help to select a synbiotic; for example, knowing the degradative enzymes that probiotics possess would help to justify selection of the combination required. Relatively little is known regarding the structure-function relationship of prebiotic oligosaccharides and their degradation in probiotics (Rastall and Maitin 2002). Mechanisms of degradation of FOS has been mostly analysed in bifidobacteria (Ehrmann *et al.* 2003; Janer *et al.* 2004; Ryan *et al.* 2005). Recent studies, using molecular techniques such as microarrays, have provided new insight into prebiotic degradation in lactobacilli (Barrangou *et al.* 2006; Goh *et al.* 2006; Saulnier *et al.* 2007).

So far, only a few combinations of probiotics/prebiotics have been evaluated as synbiotics, especially in humans, but the increasing research in this field should provide new insights into the beneficial effects of synbiotics and allow the preferred combinations to be selected for targeted groups of the population.

4. References

Akalin, A.S., Fenderya, S. and Akbulut, N. 2004. Viability and activity of bifidobacteria in yoghurts containing fructooligosaccharides during refrigerated storage. *International Journal of Food Science and Technology* **39**: 613-621.

Alander, M., Mättö, J., Kneifel, W., Johansson, M.L., Kögler, B., Crittenden, R., Mattila-Sandholm, T. and Saarela, M. 2001. Effect of galacto-oligosaccharide supplementation on human faecal microflora and on survival and persistence of *Bifidobacterium lactis* Bb-12 in the gastrointestinal tract. *International Dairy Journal* **11**: 817-825.

Anderson, A.D., McNaught, C.E., Jain, P.K. and MacFie, J. 2004. Randomised clinical trial of synbiotic therapy in elective surgical patients. *Gut* **53**: 241-245.

Aragon-Alegro, L.C., Henrique, J., Alegro, A., Cardarelli, H.R., Chiu, M.C. and Isay Saad, S.M. 2006. Potentially probiotic and synbiotic chocolate mousse. *Food Science and Technology* **40**: 669-675.

Asahara, T., Nomoto, K., Shimizu, K., Watanuki, M. and Tanaka, R. 2001. Increased resistance of mice to *Salmonella enterica* serovar Typhimurium infection by synbiotic administration of bifidobacteria and transgalactosylated oligosaccharides. *Journal of Applied Microbiology* **91**: 985-996.

Barrangou, R., Azcarate-Peril, M.A., Duong, T., Conners, S.B., Kelly, R.M. and Klaenhammer, T.R. 2006. Global analysis of carbohydrate utilization by *Lactobacillus acidophilus* using cDNA microarrays. *Proceedings of the National Academy of Sciences of the United States of America* **103**: 3816-3821.

Bartosch, S., Woodmansey, E.J., Paterson, J.C.M., McMurdo, M.E.T. and Macfarlane, G.T. 2005. Microbiological effects of consuming a synbiotic containing *Bifidobacterium bifidum*, *Bifidobacterium lactis*, and oligofructose in elderly persons, determined by real-time polymerase chain reaction and counting of viable bacteria. *Clinical Infectious Diseases* **40**: 28-37.

Bielecka, M., Biedrzycka, E. and Majkowska, A. 2002. Selection of probiotics and prebiotics for synbiotics and confirmation of their *in vivo* effectiveness. *Food Research International* **35**: 125-131.

Bouhnik, Y., Flourie, B., Andrieux, C., Bisetti, N., Briet, F. and Rambaud, J.C. 1996. Effects of *Bifidobacterium* sp. fermented milk ingested with or without inulin on colonic bifidobacteria and enzymatic activities in healthy humans. *European Journal of Clinical Nutrition* **50**: 269-273.

Branner, G.R. and Roth-Maier, D.A. 2006. Influence of pre-, pro-, and synbiotics on the intestinal availability of different B-vitamins. *Archives of Animal Nutrition* **60**: 191-204.

Bruno, F.A., Lankaputhra, W.E.V. and Shah, N.P. 2002. Growth, viability and activity of *Bifidobacterium* spp. in skim milk containing prebiotics. *Journal of Food Science* **67**: 2740-2744.

Bunout, D., Hirsch, S., de la Maza, M.P., Munoz, C., Haschke, F., Steenhout, P., Klassen, P., Barrera, G., Gattas, V. and Petermann, M. 2002. Effects of prebiotics on the immune response to vaccination in the elderly. *Journal of Parenteral and Enteral Nutrition* **26**: 372-376.

Buriti, F.C., Cardarelli, H.R. and Saad, S.M. 2007. Biopreservation by *Lactobacillus paracasei* in coculture with *Streptococcus thermophilus* in potentially probiotic and synbiotic fresh cream cheeses. *Journal of Food Protection* **70**: 228-235.

Capela, P., Hay, T.K.C. and Shah, N.P. 2006. Effect of cryoprotectants, prebiotics and microencapsulation on survival of probiotic microorganisms in yoghurts and freeze-dried yoghurt. *Food Research International* **39**: 203-211.

Cardarelli, H.R. 2006. Development of a synbiotic 'petit-suisse' cheese. PhD thesis, Faculdade de Ciências Farmacêuticas, Brasil.

Chermesh, I., Tamir, A., Reshef, R., Chowers, Y., Suissa, A., Katz, D., Gelber, M., Halpern, Z., Bengmark, S. and Eliakim, R. 2007. Failure of synbiotic 2000 to prevent postoperative recurrence of Crohn's disease. *Digestive Diseases and Sciences* **52**: 385-389.

Corcoran, B.M., Ross, R.P., Fitzgerald, G.F. and Stanton, C. 2004. Comparative survival of probiotic lactobacilli spray-dried in the presence of prebiotic substances. *Journal of Applied Microbiology* **96**: 1024-1039.

Crittenden, R., Weerakkody, R., Sanguansri, L. and Augustin, M. 2006. Synbiotic microcapsules that enhance microbial viability during nonrefrigerated storage and gastrointestinal transit. *Applied and Environmental Microbiology* **72**: 2280-2282.

De Boever, P., Wouters, R. and Verstraete, W. 2001. Combined use of *Lactobacillus reuteri* and soygerm powder as food supplement. *Letters in Applied Microbiology* **33**: 420-424.

Desai, A.R., Powell, I.B. and Shah, N.P. 2004. Survival and activity of probiotic lactobacilli in skim milk containing prebiotics. *Journal of Food Science* **69**: M57-M60.

Ehrmann, M.A., Korakli, M. and Vogel, R.F. 2003. Identification of the gene for beta-fructofuranosidase of *Bifidobacterium lactis* DSM10140(T) and characterization of the enzyme expressed in *Escherichia coli*. *Current Microbiology* **46**: 391-397.

FAO/WHO. 2002. Working group report on drafting guidelines for the evaluation of probiotics in food. 30 April-1 May, 2002. London, Canada. Available at: ftp://ftp.fao.org/es/esn/food/wgreport2.pdf.

Femia, A.P., Luceri, C., Dolara, P., Giannini, A., Biggeri, A., Salvadori, M., Clune, Y., Collins, K.J., Paglierani, M. and Caderni, G. 2002. Antitumorigenic activity of the prebiotic inulin enriched with oligofructose in combination with the probiotics *Lactobacillus rhamnosus* and *Bifidobacterium lactis* on azoxymethane-induced colon carcinogenesis in rats. *Carcinogenesis* **23**: 1953-1960.

Fooks, L.J. and Gibson, G.R. 2003. Mixed culture fermentation studies on the effects of synbiotics on the human intestinal pathogens *Campylobacter jejuni* and *Escherichia coli*. *Anaerobe* **9**: 231-242.

Furrie, E., Macfarlane, S., Kennedy, A., Cummings, J.H., Walsh, S.V., O'neil, D.A. and Macfarlane, G.T. 2005. Synbiotic therapy (*Bifidobacterium longum*/Synergy 1) initiates resolution of inflammation in patients with active ulcerative colitis: a randomised controlled pilot trial. *Gut* **54**: 242-249.

Gallaher, D.D. and Khil, J. 1999. The effect of synbiotics on colon carcinogenesis in rats. *Journal of Nutrition* **129**: 1483S-1487S.

Gibson, G.R. and Roberfroid, M.B. 1995. Dietary modulation of the human colonic microbiota - Introducing the concept of prebiotics. *Journal of Nutrition* **125**: 1401-1412.

Gmeiner, M., Kneifel, W., Kulbe, K.D., Wouters, R., De Boever, P., Nollet, L. and Verstraete, W. 2000. Influence of a synbiotic mixture consisting of *Lactobacillus acidophilus* 74-2 and a fructooligosaccharide preparation on the microbial ecology sustained in a simulation of the human intestinal microbial ecosystem (SHIME reactor). *Applied Microbiology and Biotechnology* **53**: 219-223.

Goh, Y.J., Zhang, C., Benson, A.K., Schlegel, V., Lee, J.H. and Hutkins, R.W. 2006. Identification of a putative operon involved in fructooligosaccharide utilization by *Lactobacillus paracasei*. *Applied and Environmental Microbiology* **72**: 7518-7530.

Jain, P.K., McNaught, C.E., Anderson, A.D., MacFie, J. and Mitchell, C.J. 2004. Influence of synbiotic containing *Lactobacillus acidophilus* La5, *Bifidobacterium lactis* Bb 12, *Streptococcus thermophilus*, *Lactobacillus bulgaricus* and oligofructose on gut barrier function and sepsis in critically ill patients: a randomised controlled trial. *Clinical Nutrition* **23**: 441-445.

Janer, C., Rohr, L.M., Pelaez, C., Laloi, M., Cleusix, V., Requena, T. and Meile, L. 2004. Hydrolysis of oligofructoses by the recombinant beta-fructofuranosidase from *Bifidobacterium lactis*. *Systematic and Applied Microbiology* **27**: 279-285.

Kanamori, Y., Hashizume, K., Kitano, Y., Tanaka, Y., Morotomi, M., Yuki, N. and Tanaka, R. 2003. Anaerobic dominant flora was reconstructed by synbiotics in an infant with MRSA enteritis. *Pediatrics International* **45**: 359-362.

Kanamori, Y., Hashizume, K., Sugiyama, M., Morotomi, M. and Yuki, N. 2001. Combination therapy with *Bifidobacterium breve*, *Lactobacillus casei*, and galactooligosaccharides dramatically improved the intestinal function in a girl with short bowel syndrome – A novel synbiotics therapy for intestinal failure. *Digestive Diseases and Sciences* **46**: 2010-2016.

Kanamori, Y., Hashizume, K., Sugiyama, M., Mortomi, M., Yuki, N. and Tanaka, R. 2002. A novel synbiotic therapy dramatically improved the intestinal function of a pediatric patient with laryngotracheo-esophageal cleft (LTEC) in the intensive care unit. *Clinical Nutrition* **21**: 527-530.

Kanamori, Y., Sugiyama, M., Hashizume, K., Yuki, N., Morotomi, M. and Tanaka, R. 2004. Experience of long-term synbiotic therapy in seven short bowel patients with refractory enterocolitis. *Journal of Pediatric Surgery* **39**: 1686-1692.

Kolida, S., Tuohy, K. and Gibson, G.R. 2002. Prebiotic effects of inulin and oligofructose. *British Journal of Nutrition* **87**: S193-197.

Kolida, S., Saulnier, D.M. and Gibson, G.R. 2006. Gastrointestinal microflora: probiotics. *Advances in Applied Microbiology* **59**: 187-219.

Kotzampassi, K., Giamarellos-Bourboulis, E.J., Voudouris, A., Kazamias, P. and Eleftheriadis, E. 2006. Benefits of a synbiotic formula (Synbiotic 2000Forte) in critically ill trauma patients: early results of a randomized controlled trial. *World Journal of Surgery* **30**: 1848-1855.

Lamoureux, L., Roy, D. and Gauthier, S.F. 2002. Production of oligosaccharides in yogurt containing bifidobacteria and yogurt cultures. *Journal of Dairy Science* **85**: 1058-1069.

Lesniewska, V., Rowland, I., Cani, P.D., Neyrinck, A.M., Delzenne, N.M. and Naughton, P.J. 2006. Effect on components of the intestinal microflora and plasma neuropeptide levels of feeding *Lactobacillus delbrueckii*, *Bifidobacterium lactis*, and inulin to adult and elderly rats. *Applied and Environmental Microbiology* **72**: 6533-6538.

Lilly, D.M. and Stiwell, R.H. 1965. Probiotics: Growth promoting factors produced by microorganisms. *Science* **147**: 747-748.

Liong, M.T. and Shah, N.P. 2005. Acid and bile tolerance and cholesterol removal ability of lactobacilli strains. *Journal of Dairy Science* **88**: 55-66.

Liu, Q., Duan, Z.P., Ha, D.K., Bengmark, S., Kurtovic, J. and Riordan, S.M. 2004. Synbiotic modulation of gut flora: Effect on minimal hepatic encephalopathy in patients with cirrhosis. *Hepatology* **39**: 1441-1449.

Makras, L., Van Acker, G. and De Vuyst, L. 2005. *Lactobacillus paracasei* subsp *paracasei* 8700: 2 degrades inulin-type fructans exhibiting different degrees of polymerization. *Applied and Environmental Microbiology* **71**: 6531-6537.

Malinen, E., Mättö, J., Salmitie, M., Alander, M., Saarela, M. and Palva, A. 1992. PCR-ELISA II: analysis of *Bifidobacterium* populations in human faecal samples from a consumption trial with *Bifidobacterium lactis* Bb-12 and a galacto-oligosaccharide preparation. *Systematic and Applied Microbiology* **25**: 249-258.

Morelli, L., Zonenschain, D., Callegari, M.L., Grossi, E., Maisano, F. and Fusillo, M. 2003. Assessment of a new synbiotic preparation in healthy volunteers: survival, persistence of probiotic strains and its effect on the indigenous flora. *Nutrition Journal* **9**: 2-11.

Ogawa, T., Hashikawa, S., Asai, Y., Sakamoto, H., Yasuda, K. and Makimura, Y. 2006. A new synbiotic, *Lactobacillus casei* subsp. casei together with dextran, reduces murine and human allergic reaction. *FEMS Immunology and Medical Microbiology* **46**: 400-409.

Osman, N., Adawi, D., Molin, G., Ahrne, S., Berggren, A. and Jeppsson, B. 2006. *Bifidobacterium infantis* strains with and without a combination of oligofructose and inulin (OFI) attenuate inflammation in DSS-induced colitis in rats. *BMC Gastroenterology* **6**: 31.

Ozer, D., Akin, S. and Ozer, B. 2005. Effect of inulin on survival of *Lactobacillus acidophilus* La-5 and *Bifidobacterium bifidum* BB-02 in acidophilus-bifidus yogurt. *Food Research International* **11**: 19-24.

Passeron, T., Lacour, J.P., Fontas, E. and Ortonne, J.P. 2006. Prebiotics and synbiotics: two promising approaches for the treatment of atopic dermatitis in children above 2 years. *Allergy* **61**: 431-437.

Pereira, D.I., McCartney, A.L. and Gibson, G.R. 2003. An in vitro study of the probiotic potential of a bile-salt-hydrolyzing *Lactobacillus fermentum* strain, and determination of its cholesterol-lowering properties. *Applied and Environmental Microbiology* **69**: 4743-4752.

Perrin, S., Grill, J.P. and Schneider, F. 2000. Effects of fructooligosaccharides and their monomeric components on bile salt resistance in three species of bifidobacteria. *Journal of Applied Microbiology* **88**: 968-974.

Rafter, J. et al. 2007. Dietary synbiotics reduce cancer risk factors in polypectomized and colon cancer patients. *American journal of clinical nutrition* **82**: 488-496.

Rastall, R.A. and Maitin, V. 2002. Prebiotics and synbiotics: towards the next generation. *Current Opinion in Biotechnology* **13**: 490-496.

Rayes, N., Seehofer, D., Theruvath, T., Schiller, R.A., Langrehr, J.M., Jonas, S., Bengmark, S. and Neuhaus, P. 2005. Supply of pre- and probiotics reduces bacterial infection rates after liver transplantation – A randomized, double-blind trial. *American Journal of Transplantation* **5**: 125-130.

Roller, M., Clune, Y., Collins, K., Rechkemmer, G. and Watzl, B. 2007. Consumption of prebiotic inulin enriched with oligofructose in combination with the probiotics *Lactobacillus rhamnosus* and *Bifidobacterium lactis* has minor effects on selected immune parameters in polypectomised and colon cancer patients. *British Journal of Nutrition* **97**: 676-684.

Roller, M., Pietro Femia, A., Caderni, G., Rechkemmer, G. and Watzl, B. 2004a. Intestinal immunity of rats with colon cancer is modulated by oligofructose-enriched inulin combined with *Lactobacillus rhamnosus* and *Bifidobacterium lactis*. *British Journal of Nutrition* **92**: 931-938.

Roller, M., Rechkemmer, G. and Watzl, B. 2004b. Prebiotic inulin enriched with oligofructose in combination with the probiotics *Lactobacillus rhamnosus* and *Bifidobacterium lactis* modulates intestinal immune functions in rats. *Journal of Nutrition* **134**: 153-156.

Rowland, I.R., Rumney, C.J., Coutts, J.T. and Lievense, L.C. 1998. Effect of *Bifidobacterium longum* and inulin on gut bacterial metabolism and carcinogen-induced aberrant crypt foci in rats. *Carcinogenesis* **19**: 281-285.

RTS. 2005. The European market for probiotics and prebiotics. *Ingredients Directory*.

Ryan, S.M., Fitzgerald, G.F. and van Sinderen, D. 2005. Transcriptional regulation and characterization of a novel beta-fructofuranosidase-encoding gene from *Bifidobacterium breve* UCC2003. *Applied and Environmental Microbiology* **71**: 3475-3482.

Samona, A. and Robinson, R.K. 1994. Effect of yoghurt culture on the survival of bifidobacteria in fermented milk. *Journal of Society of Dairy Technology* **47**: 58-60.

Saulnier, D.M.A. 2006. Development of a synbiotic targeted at the healthy adult population. PhD thesis, University of Reading, United Kingdom..

Saulnier, D.M.A., Molenaar, D., de Vos, W.M., Gibson, G.R. and Kolida, S. 2007. Identification of prebiotic fructooligosaccharide metabolism in *Lactobacillus plantarum* WCFS1 through microarrays. *Applied and Environmental Microbiology* **73**: 1753-1765.

Schrezenmeir, J. and de Vrese, M. 2001. Probiotics, prebiotics, and synbiotics—approaching a definition. *American Journal of Clinical Nutrition* **73**: 361S-364S.

Schultz, M.K.M., Tannock, G.W., Melchner, I., Gottl, C., Schwietz, H., Scholmerich, J. and Rath, H.C. 2004. Effects of feeding a probiotic preparation (SIM) containing inulin on the severity of colitis and on the composition of the intestinal microflora in HLA-B27 transgenic rats. *Clinical and Diagnostic Laboratory Immunology* **11**: 581-587.

Searby, L. 2006. The challenges of synbiotics. *Functional Food and Nutraceuticals* September issue.

Shin, H.S., Lee, J.H., Pestka, J.J. and Ustunol, Z. 2000. Growth and viability of commercial *Bifidobacterium* spp in skim milk containing oligosaccharides and inulin. *Journal of Food Science* **65**: 884-887.

Sugawara, G., Nagino, M., Nishio, H., Ebata, T., Takagi, K., Asahara, T., Nomoto, K. and Nimura, Y. 2006. Perioperative synbiotic treatment to prevent postoperative infectious complications in biliary cancer surgery: a randomized controlled trial. *Annals of Surgery* **244**: 706-714.

Tzortzis, G., Baillon, M.-L.A., Gibson, G.R. and Rastall, R.A. 2004. Modulation of anti-pathogenic activity in canine-derived *Lactobacillus* species by carbohydrate growth substrate. *Journal of Applied Microbiology* **96**: 552-559.

Uchida, K., Takahashi, T., Inoue, M., Morotomi, M., Otake, K., Nakazawa, M., Tsukamoto, Y., Miki, C. and Kusunoki, M. 2007. Immunonutritional effects during synbiotics therapy in pediatric patients with short bowel syndrome. *Pediatric Surgery International* **23**: 243-248.

Vanhoutte, T., De Preter, V., De Brandt, E., Verbeke, K., Swings, J. and Huys, G. 2006. Molecular monitoring of the fecal microbiota of healthy human subjects during administration of lactulose and *Saccharomyces boulardii*. *Applied and Environmental Microbiology* **72**: 5990-5997.

Van Loo, J., Clune, Y., Bennett, M. and Collins, J.K. 2005. The SYNCAN project: goals, set-up, first results and settings of the human intervention study. *British Journal of Nutrition* **93**: S91-98.

Vernazza, C.L., Gibson, G.R. and Rastall, R.A. 2006. Carbohydrate preference, acid tolerance and bile tolerance in five strains of *Bifidobacterium*. *Journal of Applied Microbiology* **100**: 845-854.

Voudouris, A., Kazamias, P., Spyridaki, E., Antonopoulou, A., Giamarellos-Bourboulis, E., Skourtis, C. and Kotzampassi, K. 2005. Benefits of Synbiotic 2000 Forte in critically ill patients: a randomized controlled trial. *Critical Care* **9**: P362.

Wang, X., Brown, I.L., Evans, A.J. and Conway, P.L. 1999. The protective effects of high amylose maize (amylomaize) starch granules on the survival of Bifidobacterium spp. in the mouse intestinal tract. *Journal of Applied Microbiology* **87**: 631-639.

Yadav, H., Jain, S. and Sinha, P.R. 2007. Formation of oligosaccharides in skim milk fermented with mixed dahi cultures, *Lactococcus lactis* ssp *diacetylactis* and probiotic strains of lactobacilli. *Journal of Dairy Research*, in press.

Bacterial spore formers as probiotics for poultry

Stephen T. Cartman[1], Roberto M. La Ragione[2] and Martin J. Woodward[2]*

[1]*Institute of Infection, Immunity and Inflammation, Centre for Biomolecular Sciences, University of Nottingham, Nottingham NG7 2RD, UK.*

[2]*Department of Food and Environmental Safety, Veterinary Laboratories Agency, Woodham Lane, New Haw, Addlestone KT15 3NB, UK. Tel. +44 (0)1932 357582. Fax +44 (0)1932 347046. E-mail m.j.woodward@vla.defra.gsi.gov.uk*

**Corresponding author*

Abstract

Traditionally, antibiotic feed supplements have been administered to intensively farmed poultry for their growth-promoting and prophylactic properties. However, their use was banned within the European Union in 2006 due to concerns over the increased prevalence of bacterial pathogens harbouring antibiotic resistance determinants. This move prompted efforts to find and develop alternative growth-promoting and prophylactic agents for poultry. Functional foods and feed supplements, including prebiotics, probiotics and synbiotics, have formed the focus of many research efforts as they may be mass-administered to poultry cost-effectively. Among the probiotic preparations currently available for poultry are a number of products comprised of bacterial spores, principally of the genus *Bacillus*. Bacterial spores are extremely robust – a highly desirable trait for commercial probiotic products as this ensures that they remain viable during distribution, storage and administration as 'live microbial products'. Furthermore, they are able to pass through the acidic gastric environment of the target host species and enter the small intestine in a viable state. To date, a limited number of published studies have demonstrated the efficacy of bacterial spores as prophylactic and performance-enhancing agents for poultry. However, a further period of study is required to elucidate the specific mechanisms of action, so that their full potential may be realised. Bacterial spores have also been successfully developed for use as oral vaccine vectors, raising the exciting prospect that a combined spore probiotic–immunoprophylactic agent may be developed sometime in the future for use in poultry.

Keywords: poultry, animals, *Bacillus*, spore, probiotics, feed supplements, prophylaxis

1. Introduction

Broilers are chickens that are reared for meat. During the rearing period they are exposed to physical and immunological stresses that leave them susceptible to infection. In addition, stocking densities inside broiler houses are typically high, with flock sizes often in the region of thousands to tens of thousands of birds. This means that outbreaks of infectious disease are able to spread rapidly throughout the flock. Therefore, in an attempt to minimise the risk of disease outbreaks, strict hygiene and biosecurity measures are enforced within hatcheries and broiler houses.

Traditionally, antibiotic feed supplements have been administered to broilers because of their prophylactic and growth-promoting properties (Bunyan *et al.* 1977). However, this practice was banned within the European Union (EU) at the beginning of 2006 (SSC 1999). This move formed part of an initiative to promote prudent antibiotic usage, in light of the increasing prevalence of bacterial pathogens that harbour antibiotic resistance determinants.

The specific mechanisms of action through which antibiotic feed supplements function remain largely unresolved; however, it has long been recognised that they act on the gastrointestinal (GI) microflora. This is exemplified by studies that have demonstrated that the growth rates of germfree chickens remain unaffected by antibiotic feed supplements, whereas the growth rates of conventionally reared chickens are enhanced by antibiotic feed supplements (Coates *et al.* 1963; Forbes and Pank 1959). It is widely accepted that a healthy and stable GI microflora protects the host against enteric infection by out-competing pathogenic microorganisms for occupation of ecological niches that exist within the GI tract (GIT). This concept is often referred to as competitive exclusion (CE). Perversely, the strict hygiene and biosecurity measures

Food Science and Technology Bulletin: Functional Foods **4** (3) 21–30
DOI: 10.1616/1476-2137.14897. Published 10 August 2007
ISSN 1476-2137

enforced within hatcheries and broiler houses are actually thought to hinder the ability of chicks to acquire a stable GI microflora, thus leaving them particularly susceptible to enteric infections. Mindful of this, the search for alternative prophylactic and growth-promoting agents has focussed on products that modulate the GI microflora favourably. Products that can be mass administered (e.g. in feed or drinking water, or by aerosol) are preferred, due to the large scale of modern broiler farms. Consequently, functional foods, including prebiotics, probiotics and synbiotics, have become the focus of many research efforts.

A prebiotic is a "food ingredient which beneficially affects the host by selectively stimulating growth, activity, or both, of one or a limited number of bacterial species present in the colon" (Gibson and Roberfroid 1995). A probiotic is often quoted as being "a live microbial feed supplement which beneficially affects the host animal by improving its intestinal balance" (Fuller 1989). This definition has now been updated to "a preparation of viable microorganisms which is consumed by humans or other animals, with the aim of inducing beneficial effects, by either qualitatively or quantitatively influencing their gut microflora and/or modifying their immune status" (Fuller 2004).

Probiotics may be divided into those which are 'defined' and those which are 'undefined', in terms of the constituent bacterial genera and species. Undefined preparations usually consist of faecal or caecal preparations from the target species of interest, whereas defined preparations usually contain one, or a select number of characterised bacterial species. Undefined preparations have been shown to be more effective as prophylactic agents (Stavric 1992; Stavric and D'Aoust 1993). However, a probiotic must be defined in order to comply with the current EU regulations on animal feed supplements (see Section 7 of this review)—a regulatory requirement aimed at ensuring the quality and safety of probiotic products (SCAN 2001a). Defined probiotics may be further categorised into those which are natural members of the GI microflora and those which are not. It has been suggested that the GIT of the target host species is the most appropriate environment from which to isolate a novel probiotic (Gibson and Fuller 2000; O'Sullivan 2001). Indeed, it follows that microorganisms isolated from the GIT of an animal will be adapted for survival and persistence within this environment.

Lactic acid bacteria, particularly lactobacilli, have arguably been the focus of most poultry probiotic studies conducted to date. This may be, at least in part, because lactobacilli predominate in the GIT of young chicks (Amit-Romach *et al.* 2004; Lu *et al.* 2003; Wise and Siragusa 2007). Furthermore, a number of species are facultative anaerobes. Therefore, while they are well-adapted to the anoxic conditions in the GI environment, they will not lose viability when exposed to atmospheric oxygen during distribution, storage and administration as a 'live microbial product'.

Spore-forming bacteria of the genus *Bacillus* have been studied as probiotics (reviewed in Cartman and La Ragione 2004; Hong *et al.* 2005; Mazza 1994; Sanders *et al.* 2003; Senesi 2004). *Bacillus* spp. are often considered to be soil inhabitants. However, species of this genus are ubiquitous and, therefore, the places in which they are found do not necessarily indicate their natural habitat (Nicholson 2002, 2004). Although *Bacillus* spp. do not colonise the GIT of man or animals, they may be isolated from GI samples (Barbosa *et al.* 2005). Furthermore, a body of evidence is emerging to suggest that *Bacillus* spp. are able to survive and replicate within the GIT of man and animals (see Section 5 of this review).

This review will focus exclusively on the use of spore-forming bacteria as probiotics for poultry, with mention of studies conducted in other animal species where deemed appropriate. For a comprehensive review of the scientific literature relating to the use of bacterial spore formers as probiotics for humans and animals other than poultry, the reader is referred to four excellent articles in this field (Hong *et al.* 2005; Mazza 1994; Sanders *et al.* 2003; Senesi 2004).

2. Bacterial spore formers

Most species of spore-forming bacteria belong to one of the two principle genera, *Bacillus* (aerobic or facultative aerobic species) or *Clostridium* (anaerobic species), although other genera do exist (e.g. *Brevibacillus* and *Sporolactobacillus*; see Fritze (2004) for a comprehensive review on the taxonomy of spore-forming bacteria). At present, the field of spore probiotics is almost exclusively concerned with the use of *Bacillus* species. Moreover, at the present time, we are unaware of any defined probiotic product that is knowingly comprised of, or contains, species of the genus *Clostridium*.

Species of the genus *Bacillus* are characterised as rod shaped, aerobic or facultative aerobic, Gram-positive, spore-forming bacteria. *Bacillus subtilis* is arguably the most extensively characterised species of the genus and is a paradigm for studies on Gram-positive bacteria, particularly with regard to its ability to undergo cellular differentiation (i.e. sporulate and germinate). Vegetative cells are the metabolically active cell form, which grow and replicate when environmental conditions permit. When conditions become unfavourable for cell growth and replication, an alternative programme of cell division occurs in which vegetative cells produce spores as a means of spatial and temporal escape (Nicholson *et al.* 2000). Spores are metabolically dormant and extremely resilient to environmental stresses, including extremes of temperature and pressure, and exposure to ultraviolet radiation and noxious chemicals

(Nicholson *et al.* 2000; Nicholson 2004). Indeed, bacterial spores have been documented as having survived intact for 28 million years (Cano and Borucki 1995). When conditions become favourable for cell growth and replication again, spores may germinate and re-enter the vegetative cell cycle. For a comprehensive review on sporulation in *B. subtilis* the reader is referred to an excellent review by Stragier and Losick (1996). Likewise, for comprehensive reviews on spore germination, the reader is referred to two excellent articles by Moir *et al.* (2002) and Setlow (2003).

3. Bacterial spores as live microbial products

A principle consideration for any probiotic manufacturer is to optimise the number of viable (probiotic) cells which enter the GIT of the target host species. With this in mind, a major challenge that must be addressed is the preservation of cell viability during the distribution, storage and administration of a probiotic product. The intrinsic robustness of bacterial spores means that they are ideally suited to this commercial consideration without the need for refrigeration or specialist packaging to protect them from changeable environmental conditions. Furthermore, because bacterial spores are metabolically dormant, there is no need for them to be packaged in a nutrient-rich storage matrix to preserve cell viability. Overall, this means that probiotic products comprised of bacterial spores have a long shelf life when compared to products which are comprised of microorganisms with more fragile cell forms. It is also noteworthy that the robustness of bacterial spores enables them to pass through the acidic gastric barrier (i.e. stomach) of the target host species and enter the small intestine in a viable state (Duc *et al.* 2003a, 2004a).

4. Efficacy of spore probiotics

A limited number of studies have been published in the scientific literature providing evidence that bacterial spore formers are efficacious as prophylactic and/or performance-enhancing agents. With regards to prophylaxis, two different strains of *B. subtilis* have been shown to facilitate the exclusion of *Clostridium perfringens* (the aetiological agent of necrotic enteritis in poultry and fowl), avian pathogenic *Escherichia coli* (APEC) and *Salmonella enterica* serovar Enteritidis from the GIT of poultry (La Ragione *et al.* 2001; La Ragione and Woodward 2003; Maruta *et al.* 1996). In addition, one of these strains has also been shown to facilitate the exclusion of *Campylobacter jejuni* (Maruta *et al.* 1996). A strain of *B. coagulans* has also been shown to have both a prophylactic and a growth-promoting effect on broilers, which was comparable to that of the antibiotic feed supplement virginiamycin (Cavazzoni *et al.* 1998).

In general, studies on the performance-enhancing properties of spore probiotics have focussed on their ability to enhance feed conversion efficiencies (i.e. weight of feed consumed per weight gained by an animal, over a fixed time period). To this end, improved feed conversion efficiencies have been documented in at least three studies in which chickens were fed preparations of *B. subtilis* (Jiraphocakul *et al.* 1990; Samanya and Yamauchi 2002; Santoso *et al.* 1995). Furthermore, in one study it was noted that chickens fed *B. subtilis* had a reduced blood ammonia concentration which, interestingly, correlated with enhanced intestinal function, as measured by villus height and enterocyte proliferation (Samanya and Yamauchi 2002). This might indicate one mechanism through which bacterial spore formers may function as probiotics.

Work conducted in our laboratory suggests that the benefits conferred by spore probiotics are strain specific (Cartman 2006). Although the number of studies conducted to date is limited, they clearly indicate that certain strains, of spore-forming bacteria at least, harbour properties that make them efficacious as probiotics for poultry. However, it is apparent that further studies are necessary in order to identify exactly what these properties are and, ultimately, realise the full potential of bacterial spore formers as prophylactic and/or performance-enhancing agents. To do this, it is likely that more empirical studies, incorporating many more strains of spore-forming bacteria, will be required in order to establish a well-rounded perspective on the overall value of bacterial spore formers as probiotics for poultry.

5. The fate of ingested spores

In order to begin to understand exactly how bacterial spores may function as probiotics, it is necessary to consider what happens to them once they have entered the GIT of the target host species. It is noteworthy that the definition of a probiotic (see Introduction in this review) stipulates that a probiotic should be 'live' or 'viable'. This implies that probiotic metabolism within the host GIT is a necessary prerequisite for any beneficial effect to occur. With this in mind, it is worth considering that although bacterial spores are 'viable', they are metabolically dormant cell forms. This raises the important question of whether or not orally-administered spores are able to germinate in the gut and, subsequently, function as probiotics in the host GIT as vegetative cells.

The first studies that indicated that spores of *Bacillus* spp. could germinate in the GIT were conducted in rabbits and dogs (Hisanga 1980). In more recent times, comprehensive *in vivo* studies have been conducted in mice. Initially, it was thought that bacterial spores were not able to germinate in the GIT of mice (Spinosa *et al.* 2000). However, more recently, it was observed that, on occasion, a greater number of spores could be recovered from mouse faecal material than were present in the original oral dose of *B. subtilis* administered (Hoa *et al.* 2001). It was hypothesised that the spores must have germinated, replicated and resporulated in the GIT of the mice. This finding was reproduced in a follow-up study and a molecular approach was used to prove conclusively that spores of *B. subtilis* are able to germinate in the small intestine of mice (Casula and Cutting 2002). Since this work, other studies have also demonstrated bacterial spore germination in the GIT of mice (Duc *et al.* 2003a, 2004b).

The germination of bacterial spores in the GIT of poultry has only been investigated in one study, in which it was found that spores of *B. cereus* var. *toyoi* germinate rapidly (i.e. within hours of administration) in the GIT of chickens (Jadamus *et al.* 2001). It was also concluded that spore germination was a necessary prerequisite for any probiotic (i.e. beneficial) effect to occur. Recently, work conducted in our laboratory has found that spores of *B. subtilis* are also able to germinate in the GIT of chickens, giving rise to metabolically active vegetative cells within this host (Cartman 2006).

Bacillus spp. are generally regarded as being aerobic bacteria. Therefore, it is necessary to consider if, and how, vegetative cells may survive in the anoxic conditions of the GIT, following spore germination. Interestingly, the model organism *B. subtilis* is able to grow and replicate under anaerobic conditions, either by fermentation, or by using nitrite or nitrate as a terminal electron acceptor in the electron transport chain (Nakano *et al.* 1997; Nakano and Zuber 1998). This suggests that *Bacillus* spp. *are* able to survive in the anoxic conditions of the GIT. Indeed, it has now been demonstrated that spores of *B. subtilis* may undergo a full life cycle in the GIT of mice (i.e. germination, vegetative cell growth and resporulation) before being egested (Tam *et al.* 2006).

Although the studies mentioned so far indicate that orally administered spores can undergo a full life cycle in the GIT, it is noteworthy that *Bacillus* spp. are considered to be transient members of the GI microflora as they do not 'colonise' the guts of man and animals. Notwithstanding this, it has been found in work conducted in our laboratory that *B. subtilis* can be detected in the GIT of chickens for up to six weeks after an oral dose of spores has been administered (Cartman 2006). Considering that it has been estimated to take approximately six to seven hours, following ingestion, for digesta to pass through the GIT of a chicken (Shires *et al.* 1987), this finding suggests that *B. subtilis* is not merely an inert passenger in the GIT of this host.

6. Mechanisms of action

Although bacterial spores do not appear to be inert passengers in the GIT of man and animals, our knowledge of how they may function as probiotics is severely limited. Indeed, much of our knowledge regarding the mechanisms of action through which probiotics, in general, may function remains conjecture at present. The diversity and dynamic nature of microorganisms in the GI environment of animals and man makes gut microbiology an extremely complex area of research. However, in principle, the mechanisms of action through which bacterial spores may function as probiotics are the same as those for any probiotic microorganism.

In broiler farming, a probiotic may be used as a prophylactic agent and/or a performance-enhancing agent. In both cases the probiotic may act directly on the host (e.g. stimulate the immune system), alter the dynamics of the existing GI-microflora (which in turn may enhance feed conversion efficiency and/or enhance the ability of the microflora to exclude one or more pathogens), or it may carry out a function in the gut lumen that is beneficial to the host (e.g. enhance digestion or nutrient availability by breaking down otherwise indigestible food substrates).

Specific mechanisms of action through which bacterial spores may function as prophylactic and/or performance-enhancing agents in poultry are listed in Tables 1 and 2, respectively. It should be noted that the mechanisms of action listed are *not* mutually exclusive and a probiotic may function through any one, or a combination of the mechanisms listed. Whether or not Tables 1 and 2 provide a comprehensive summary of how spore probiotics actually function in the avian GIT remains unclear as the mechanisms of action listed represent active areas of research, many of which are still at an early stage.

7. Regulation

7.1 The European perspective

Animal feed additives (including microorganisms) are closely regulated within the EU. The ban on the use of antimicrobial feed supplements in 2006 prompted an enhanced interest in alternative feed supplements and, as a result, the legislation governing this area has undergone significant change in recent years. Until May 2003, the risk assessment of animal feed additives for use in Europe was the responsibility of the Scientific Committee of Animal Nutrition (SCAN; Anadon *et al.* 2006; von Wright 2005). After this date, SCAN was superseded by the European Food Safety Authority (EFSA). While the EFSA provide expert scientific advice to the European Commission (EC), ultimately, the approval and risk management of a probiotic product is the responsibility of the EC and its constituent member states. Approval of a novel product requires a favourable vote from a qualified majority of member states and, if successful, a license is granted for an initial period of four years (McCartney 2004).

Table 1. Mechanisms of action through which spore probiotics may function as prophylactic agents in the gastrointestinal tract (GIT) of a chicken

Mechanism of probiotic action	Target of probiotic action	Is spore germination (giving rise to metabolically active vegetative cells) in the GIT necessary?
Stimulation of the host immune system	Host organism	Dependent on whether or not vegetative cell form is required for probiotic to function via this mechanism
Competition for host cell receptor sites	Microorganisms in the host GIT	Dependent on whether or not vegetative cell form is required for probiotic to function via this mechanism
Coaggregation with pathogenic microorganisms (inhibiting their ability to adhere to and colonise the gut mucosa)	Microorganisms in the host GIT	Dependent on whether or not vegetative cell form is required for probiotic to function via this mechanism
Secretion of antimicrobial and/or inhibitory molecules	Microorganisms in the host GIT	Yes
Competition for essential nutrients	Microorganisms in the host GIT	Yes

Much of the current legislation on animal feed additives (including microorganisms) has been derived from opinions and guidelines published by the SCAN or EFSA. Before 18th October 2004, animal feed additives were regulated by EC Council Directive 70/524/EEC and its numerous amendments, while Council Directive 87/153/EEC and its amendments defined the guidelines for the assessment of feed additives. Both of these have now been replaced by EC Regulation 1831/2003 (Anadon *et al.* 2006; von Wright 2005). Perhaps the most significant difference since the implementation of this new regulation is that member states can now consider non-scientific criteria when assessing a product. Therefore, even if the EFSA issue a favourable (scientific) opinion on a novel product, a member state may decide not to approve it based on social, economic or ethical reasons. The current EFSA assessment criteria for novel microbial feed additives are summarised in Table 3 (EFSA 2005c; SCAN 2001a, 2003).

Because certain *Bacillus* spp. are able to produce toxins, the SCAN issued an opinion specifically to address the use of *Bacillus* spp. in animal nutrition (SCAN 2000a). *B. cereus* was a major concern as this species has the potential to produce haemolysins, diarrhoeal enterotoxins and an emetic toxin. Moreover, some strains of *B. cereus* were available as probiotics but had not been subject to rigorous testing, with regard to their toxin-producing potential (Green *et al.* 1999; Hoa *et al.* 2000). Consequently, the SCAN recommended that, for all applications involving the addition of living organisms to animal feed, the use of strains belonging to the *B. cereus* taxonomic group should be strongly discouraged (SCAN 2000a). Furthermore, commercial toxin testing assays, toxin bioassays and molecular (i.e. PCR-based) toxin tests must now be carried out on all *Bacillus* spp., for which EU approval is being sought (Anadon *et al.* 2006).

7.2 The United States' perspective

According to the United States Federal Food, Drug and Cosmetics Act, a microorganism that is used as a feed additive is subject to approval by the Food and Drug Administration (FDA) on the basis of safety and efficacy data, unless it is generally recognised as safe (GRAS). GRAS status may be attained in one of two ways: either a microorganism has a history of safe use in food (dating

Table 2. Mechanisms of action through which spore probiotics may function as a performance-enhancing agents in the gastrointestinal tract (GIT) of a chicken (i.e. enhance feed conversion efficiency)

Mechanism of probiotic action	Target of probiotic action	Is spore germination (to yield metabolically active vegetative cells) in the GIT necessary?
Enhance digestion/nutrient availability (by breaking down substrates in gut lumen)	Food substrates in the gut lumen	Yes
Stimulate intestinal function (e.g. villus height and enterocyte proliferation)	Host organism	Dependent on whether or not vegetative cell form is required for probiotic to function via this mechanism

Table 3. Summary of EU regulatory assessment criteria for microbial feed additives*

Assessment category	Category components (information must be provided where applicable)
Identity of the additive	• Proprietary name • Type of additive according to its main function • Qualitative and quantitative composition (of both active ingredient and impurities) • Physical state of each form of the product • Summary of manufacturing process
Characterisation of the active agent(s)	• Nomenclature • Biological origin (including origin of bacterial strain(s), culture collection deposits, morphology, physiology, molecular characteristics, taxonomy, genotypic and phenotypic stability) • Genetic modification • Compliance with release directive for genetically modified organisms (GMOs) • Toxins and virulence factors • Antibiotic production and antibiotic resistance • Other relevant properties (e.g. components of growth medium if this might be incorporated into final product, batch to batch variation, quality control procedures and historical uses of the product prior to assessment)
Characterisation of the additive: physicochemical and technological properties	• Stability (including photostability and stability during storage) • Other physicochemical or biological properties (e.g. ease with which homogenous mixtures can be attained in premixtures and feedstuffs) • Incompatibilities with other feed ingredients
Conditions of use	• Proposed conditions of use (i.e. target animal species, dosage to be administered, restrictions on use) • Safety data sheet
Control methods	• General methods – i.e. tests used in characterisation and routine quality control procedures – may include: viable counts (of probiotic microorganism), moisture content, heavy metal content, particle size of final product, impurities (chemical and biological), dust-forming potential, compatibility with other approved feedstuffs, additives or medicinal products • Description of qualitative and quantitative methods used for routine control of active agent in premixtures and feeding stuffs
Effect(s) on animals	• Evidence of efficacy (for the claims made by the manufacturer) – supporting evidence must consist of a description and data from at least three independent trials. These should be conducted in at least two different locations and results must show statistical significance. For multicomponent additives (i.e. those comprised of two or more probiotic microorganisms), efficacy must be demonstrated for the complete feed additive at the minimum dose recommended by the manufacturer
Studies on target species	• Tolerance – target species should tolerate at least 10 times the maximum dose recommended by the manufacturer, without displaying any adverse effects • Effects on the gastrointestinal microflora – if it is claimed that a probiotic product has an effect on the host microflora, supporting data must be provided • Metabolism and residue studies – if it is claimed that a probiotic product has an effect on host metabolism, supporting data must be provided
Consumer safety assessment	• Genotoxicity studies • Oral toxicity studies
Worker safety assessment	• Irritancy (to skin and eyes) • Skin sensitisation • Toxic effects on the respiratory system • Systemic toxicity • Control measures (i.e. details of personal protective equipment required)
Environmental risk assessment	• Only required in exceptional cases (e.g. genetically modified microbial products)

*Comprehensive resources on current EU regulations available at: http://europa.eu.int/comm/food/food/animalnutrition/feedadditives/index_en.htm.

Table 4. Bacterial spore probiotic products available for poultry

Product	Constituent *Bacillus* spp. (culture collection reference)	Manufacturer	Comments/References
BioPlus 2B®	*B. subtilis* (DSM 5750) *B. licheniformis* (DSM 5749)	Christian Hansen Biosystems, Denmark	*EU approved* (SCAN 2000b) Also sold as 'BioGrow®' by Provita Eurotech Ltd, NI
Calsporin®	*B. subtilis* C-3102 (DSM 15544)	Calpis Co. Ltd., Japan	*EU approved* (Maruta *et al.* 1996; EFSA 2005c)
Neoferm BS 10®	Two strains of *B. clausii* (CNCM MA23/3V and CNCM MA66/4M)	Sanofi Sante Nutrition Animale, France	*Not EU approved* (SCAN 1999)
Paciflor® C10	*B. cereus* (CIP 5832/ATCC 14893)	Intervet International	*Not EU approved* Withdrawn from production in 2002 (SCAN 2001b)
Toyocerin®	*B. cereus* var. *toyoi* (NCIMB-40112/CNCM-1012)	Asahi Vet S.A., Japan	*EU approved* (SCAN 2001c)

Culture collection abbreviations: ATCC, American Type Culture Collection, USA; CIP, Collection de L'institute Pasteur, France; CNCM, Collection Nationale de Cultures de Microorganismes, France; DSM, Deutsche Sammlung von Mikrooganismen, Germany; NCIMB, National Collection of Industrial, Marine and Food Bacteria, UK.

back to before 1 January 1958), or it has been recognised as safe by qualified experts, under the intended conditions of use (von Wright 2005). In practice, GRAS status is a means of bypassing the FDA approval process in cases where it is considered to be excessive and/or unnecessary. In practice, no probiotic product has been granted GRAS status, although a number of *Bacillus* spp. (including *B. subtilis*) have been granted GRAS status for specific applications (e.g. enzyme production; Sanders *et al.* 2003). However, it is noteworthy that the use of antibiotic feed supplements has *not* been banned in the USA.

Interestingly, the EFSA are considering introducing legislation into the EU regulatory framework which is similar to that of the United States' GRAS status (EFSA 2005b). Termed 'Qualified Presumption of Safety' (QPS), it is proposed that this legislation will be tailored to account for regulatory practices in Europe. In practice the QPS concept will serve a similar function to the United States' GRAS status, in that it will serve as a means of bypassing the expensive and time-consuming approval process in cases where it is considered to be excessive and/or unnecessary (e.g. where a product has a proven track record in human foodstuffs).

8. Commercially available products

A number of spore probiotic products are currently available for poultry (Table 4). Only three have been granted EU approval: BioPlus 2B® (SCAN 2000b), Calsporin® (EFSA 2005a) and Toyocerin® (SCAN 2001b). Toyocerin® is an interesting case because it is comprised of a strain of *B. cereus* (var. *toyoi*), a potentially toxigenic species of the *Bacillus* genus. However, this product was granted EU approval as the manufacturer was able to demonstrate that *B. cereus* var. *toyoi* is non-toxigenic and unable to transfer antibiotic resistance determinants to other bacteria (SCAN 2001b). In contrast, Paciflor® C10 was not granted EU approval. The strain of *B. cereus* in this product was shown to produce at least two major enterotoxins associated with food poisoning in humans, therefore, the SCAN concluded that this product posed a risk to human health (SCAN 2001c). Paciflor® C10 was subsequently withdrawn from production in 2002.

Neoferm BS 10® was also not approved due to the potential for the *B. clausii* strains in this product to transfer antibiotic resistance determinants to other bacteria (SCAN 1999). The two strains of *B. clausii* in Neoferm BS 10® were originally derived from a human probiotic, intended to be administered as an adjunct to antibiotic therapy. As such, both strains are resistant to a number of clinically important antibiotics. Their potential to transfer erythromycin methylase (*erm*) genes was of particular concern to the SCAN (SCAN 1999).

9. Developing the spore probiotic concept

In addition to their use as probiotics, bacterial spores have been successfully developed as oral vaccine vectors (reviewed in Cutting 2004; Duc and Cutting 2003; Ferreira *et al.* 2005; Oggioni *et al.* 2003; Ricca and Cutting 2003). This raises the exciting prospect that a combined spore–probiotic–oral vaccine candidate could be developed for poultry. In seminal studies conducted in a murine model, protective immune responses were stimulated against the tetanus toxin of *Clostridium tetani* (Duc *et al.* 2003b; Isticato *et al.* 2001; Mauriello *et al.* 2004; Uyen *et al.* 2007) and *B. anthracis*, the causative agent of anthrax (Duc *et al.* 2007). This was achieved by orally administering recombinant spores of *B. subtilis* which had been genetically engineered to express heterologous antigen, thus engaging the host immune system via the gut-associated lymphoid tissue (GALT).

Strategies that have been used for heterologous antigen expression in *B. subtilis* include: expression of antigen on the surface of spores (Duc *et al.* 2003b; Isticato *et al.* 2001; Mauriello *et al.* 2004), expression of antigen within vegetative cells (Duc *et al.* 2003a; Uyen *et al.* 2007), secretion of antigen by vegetative cells (Duc *et al.* 2007) and combinations of these strategies. Interestingly, different antigen presentation strategies have been found to differentially stimulate the host immune system; therefore, in theory it is possible to tailor the antigen presentation strategy for the disease in question (Duc *et al.* 2007; Uyen *et al.* 2007).

An initial period of study would be required in order to assess the feasibility of using recombinant spores as oral vaccine preparations for poultry. However, it may ultimately be possible to use this approach to develop immunoprophylactic agents which, potentially, may be useful for controlling coccidial, bacterial and/or viral pathogens in commercial poultry flocks.

10. Conclusions

Bacterial spores are extremely robust cell forms. This innate property means that they are ideally suited for use as 'live microbial products', as they are able to retain viability during distribution, storage and administration. To date, a limited number of studies have been published that demonstrate the efficacy of bacterial spores as prophylactic and performance-enhancing agents for poultry. In addition, a number of spore probiotic products are already available for poultry. However, in order to harness the full potential of these products, a period of further research will be required to elucidate their specific mechanisms of action. In addition, it is likely that more empirical studies, incorporating many more strains of spore-forming bacteria, will be required in order to establish a well-rounded perspective on the overall value of bacterial spore formers as probiotics for poultry. With this knowledge, it will be possible to establish suitable selection criteria for identifying novel spore probiotics that are efficacious in poultry and, ultimately, it may be possible to 'tailor' spore probiotic products through genetic manipulation in order to optimise the benefits they confer on the host.

11. References

Amit-Romach, E., Sklan, D. and Uni, Z. 2004. Microflora ecology of the chicken intestine using 16S ribosomal DNA primers. *Poultry Science* **83**: 1093-1098.

Anadon, A., Martinez-Larranaga, M.R. and Aranzazu Martinez, M. 2006. Probiotics for animal nutrition in the European Union. Regulation and safety assessment. *Regulatory Toxicology and Pharmacology* **45**: 91-95.

Barbosa, T.M., Serra, C.R., La Ragione, R.M., Woodward, M.J. and Henriques, A.O. 2005. Screening for *Bacillus* isolates in the broiler gastrointestinal tract. *Applied and Environmental Microbiology* **71**: 968-978.

Bunyan, J., Jeffries, L., Sayers, J.R., Gulliver, A.L. and Coleman, K. 1977. Antimicrobial substances and chick growth promotion: the growth promoting activities of antimicrobial substances, including fifty-two used either in therapy or as dietary feed additives. *British Poultry Science* **18**: 283-294.

Cano, R.J. and Borucki, M.K. 1995. Revival and identification of bacterial spores in 25- to 40-million-year-old Dominican amber. *Science* **268**: 1060-1064.

Cartman, S.T. and La Ragione, R.M. 2004. Spore probiotics as animal feed supplements. In: Ricca, E., Henriques, A.O. and Cutting, S.M., editors. *Bacterial spore formers: probiotics and emerging applications:* 155-161. Horizon Bioscience, UK.

Cartman, S.T. 2006. The use of spores of *Bacillus subtilis* for prophylactic control of *Clostridium perfringens*-associated necrotic enteritis. PhD thesis, University of London, UK.

Casula, G. and Cutting, S.M. 2002. *Bacillus* probiotics: spore germination in the gastrointestinal tract. *Applied and Environmental Microbiology* **68**: 2344-2352.

Cavazzoni, V., Adami, A. and Castrovilli, C. 1998. Performance of broiler chickens supplemented with *Bacillus coagulans* as probiotic. *British Poultry Science* **39**: 526-529.

Coates, M.E., Fuller, R., Harrison, G.F., Lev, M. and Suffolk, S.F. 1963. A comparison of the growth of chicks in the Gustafsson germ-free apparatus and in a conventional environment, with and without dietary supplements of penicillin. *British Journal of Nutrition* **17**: 141-150.

Cutting, S.M. 2004. Spores as oral vaccines. In: Ricca, E., Henriques, A.O. and Cutting, S.M., editors. *Bacterial spore formers: probiotics and emerging applications:* 201-206. Horizon Bioscience, UK.

Duc L.H. and Cutting, S.M. 2003. Bacterial spores as heat stable vaccine vehicles. *Expert Opinion on Biological Therapy* **3**: 1263-1270.

Duc L.H., Hong, H.A. and Cutting, S.M. 2003a. Germination of the spore in the gastrointestinal tract provides a novel route for heterologous antigen delivery. *Vaccine* **21**: 4215-4224.

Duc L.H., Hong, H.A., Fairweather, N., Ricca, E. and Cutting, S.M. 2003b. Bacterial spores as vaccine vehicles. *Infection and Immunity* **71**: 2810-2818.

Duc L.H., Hong, H.A., Barbosa, T.M., Henriques, A.O. and Cutting, S.M. 2004a. Characterization of *Bacillus* probiotics available for human use. *Applied and Environmental Microbiology* **70**: 2161-2171.

Duc L.H., Hong, H.A., Uyen, N.Q. and Cutting, S.M. 2004b. Intracellular fate and immunogenicity of *B. subtilis* spores. *Vaccine* **22**: 1873-1885.

Duc L.H., Hong, H.A., Atkins, H.S., Flick-Smith, H.C., Durrani, Z., Rijpkema, S., Titball, R.W. and Cutting, S.M. 2007. Immunization against anthrax using *Bacillus subtilis* spores expressing the anthrax protective antigen. *Vaccine* **25**: 346-355.

EFSA. 2005a. European Food Safety Authority public summary on Calsporin. Available at: http://www.efsa.eu.int/science/feedap/an_applications/summaries/1138/calsporin_efsapublicsummaryen1.pdf.

EFSA. 2005b. Scientific Colloquium Summary Report. QPS (Qualified Presumption of Safety of Microorganisms) in Food and

Feed. Available at: http://www.efsa.eu.int/science/colloquium_series/no2_qps/948_en.html.

EFSA. 2005c. Opinion of the Scientific Panel on Additives and Products or Substances used in Animal Feed on the updating of the criteria used in the assessment of bacteria for resistance to antibiotics of human or veterinary importance. *The European Food Safety Authority Journal.* **223**: 1-12.

Ferreira, L.C., Ferreira, R.C. and Schumann, W. 2005. *Bacillus subtilis* as a tool for vaccine development: from antigen factories to delivery vectors. *Anais da Academia Brasileira de Ciencias* **77**: 113-124.

Forbes, M. and Pank, J.T. 1959. Growth of germ-free and conventional chicks: Effect of diet, dietary penicillin and bacterial environment. *Journal of Nutrition* **67**: 69-84.

Fritze, D. 2004. Taxonomy and systematics of the aerobic endospore forming bacteria: *Bacillus* and related genera. In: Ricca, E., Henriques, A.O. and Cutting, S.M., editors. *Bacterial spore formers: probiotics and emerging applications:* 17-34. Horizon Bioscience, UK.

Fuller, R. 1989. Probiotics in man and animals. *Journal of Applied Bacteriology* **66**: 365-378.

Fuller, R. 2004. What is a probiotic? *Biologist* **51**: 232.

Gibson, G.R. and Roberfroid, M.B. 1995. Dietary modulation of the human colonic microbiota: introducing the concept of prebiotics. *Journal of Nutrition* **125**: 1401-1412.

Gibson, G.R. and Fuller, R. 2000. Aspects of *in vitro* and *in vivo* research approaches directed toward identifying probiotics and prebiotics for human use. *Journal of Nutrition* **130**: 391S-395S.

Green, D.H., Wakeley, P.R., Page, A., Barnes, A., Baccigalupi, L., Ricca, E. and Cutting, S.M. 1999. Characterization of two *Bacillus* probiotics. *Applied and Environmental Microbiology* **65**: 4288-4291.

Hisanga, S. 1980. Studies on the germination of genus *Bacillus* spores in rabbit and canine intestines. *Journal of Nagoya City Medical Association* **30**: 456-469.

Hoa, N.T., Baccigalupi, L., Huxham, A., Smertenko, A., Van, P.H., Ammendola, S., Ricca, E. and Cutting, A.S. 2000. Characterization of *Bacillus* species used for oral bacteriotherapy and bacterioprophylaxis of gastrointestinal disorders. *Applied and Environmental Microbiology* **66**: 5241-5247.

Hoa, T.T., Duc L.H., Isticato, R., Baccigalupi, L., Ricca, E., Van, P.H. and Cutting, S.M. 2001. Fate and dissemination of *Bacillus subtilis* spores in a murine model. *Applied and Environmental Microbiology* **67**: 3819-3823.

Hong, H.A., Duc L.H. and Cutting, S.M. 2005. The use of bacterial spore formers as probiotics. *FEMS Microbiology Reviews* **29**: 813-835.

Isticato, R., Cangiano, G., Tran, H.T., Ciabattini, A., Medaglini, D., Oggioni, M.R., De Felice, M., Pozzi, G. and Ricca, E. 2001. Surface display of recombinant proteins on *Bacillus subtilis* spores. *Journal of Bacteriology* **183**: 6294-6301.

Jadamus, A., Vahjen, W. and Simon, O. 2001. Growth behaviour of a spore forming probiotic strain in the gastrointestinal tract of broiler chicken and piglets. *Archiv fur Tierernahrung (Berlin)* **54**: 1-17.

Jiraphocakul, S., Sullivan, T.W. and Shahani, K.M. 1990. Influence of a dried *Bacillus subtilis* culture and antibiotics on performance and intestinal microflora in turkeys. *Poultry Science* **69**: 1966-1973.

La Ragione, R.M., Casula, G., Cutting, S.M. and Woodward, M.J. 2001. *Bacillus subtilis* spores competitively exclude *Escherichia coli* O78:K80 in poultry. *Veterinary Microbiology* **79**: 133-142.

La Ragione, R.M. and Woodward, M.J. 2003. Competitive exclusion by *Bacillus subtilis* spores of *Salmonella enterica* serotype Enteritidis and *Clostridium perfringens* in young chickens. *Veterinary Microbiology* **94**: 245-256.

Lu, J., Idris, U., Harmon, B., Hofacre, C., Maurer, J.J. and Lee, M.D. 2003. Diversity and succession of the intestinal bacterial community of the maturing broiler chicken. *Applied and Environmental Microbiology* **69**: 6816-6824.

Maruta, K., Miyazaki, H., Masuda, S., Takahashi, M., Marubashi, T., Tadano, Y. and Takahashi, H. 1996. Exclusion of intestinal pathogens by continuous feeding with *Bacillus subtilis* C-3102 and its influence on the intestinal microflora in broilers. *Animal Science and Technology (Jpn)* **67**: 273-280.

Mauriello, E.M., Duc L.H., Isticato, R., Cangiano, G., Hong, H.A., De Felice, M., Ricca, E. and Cutting, S.M. 2004. Display of heterologous antigens on the *Bacillus subtilis* spore coat using CotC as a fusion partner. *Vaccine* **22**: 1177-1187.

Mazza, P. 1994. The use of *Bacillus subtilis* as an antidiarrhoeal microorganism. *Bollentio Chimico Farmaceutico* **133**: 3-18.

McCartney, E. 2004. EU regulations on Bacillary probiotics for animal feeds. In: Ricca, E., Henriques, A.O. and Cutting, S.M., editors. *Bacterial spore formers: probiotics and emerging applications:* 221-227. Horizon Bioscience, UK.

Moir, A., Corfe, B.M. and Behravan, J. 2002. Spore germination. *Cellular and Molecular Life Sciences* **59**: 403-409.

Nakano, M.M., Dailly, Y.P., Zuber, P. and Clark, D.P. 1997. Characterization of anaerobic fermentative growth of *Bacillus subtilis*: identification of fermentation end products and genes required for growth. *Journal of Bacteriology* **179**: 6749-6755.

Nakano, M.M. and Zuber, P. 1998. Anaerobic growth of a "strict aerobe" (*Bacillus subtilis*). *Annual Reviews in Microbiology* **52**: 165-190.

Nicholson, W.L., Munakata, N., Horneck, G., Melosh, H.J. and Setlow, P. 2000. Resistance of *Bacillus* endospores to extreme terrestrial and extraterrestrial environments. *Microbiology and Molecular Biology Reviews* **64**: 548-572.

Nicholson, W.L. 2002. Roles of *Bacillus* endospores in the environment. *Cellular and Molecular Life Sciences* **59**: 410-416.

Nicholson, W.L. 2004. Ubiquity, longevity and ecological roles of *Bacillus* spores. In: Ricca, E., Henriques, A.O. and Cutting, S.M., editors. *Bacterial spore formers: probiotics and emerging applications:* 1–15. Horizon Bioscience, UK.

O'Sullivan, D.J. 2001. Screening of intestinal microflora for effective probiotic bacteria. *Journal of Agricultural Food Chemistry* **49**: 1751-1760.

Oggioni, M.R., Ciabattini, A., Cuppone, A.M. and Pozzi, G. 2003. *Bacillus* spores for vaccine delivery. *Vaccine* **21** (Supplement 2) S96-101.

Ricca, E. and Cutting, S.M. 2003. Emerging applications of bacterial spores in nanobiotechnology. *Journal of Nanobiotechnology* **1**: 6.

Samanya, M. and Yamauchi, K.E. 2002. Histological alterations of intestinal villi in chickens fed dried *Bacillus subtilis* var. *natto*. *Comparative Biochemistry and Physiology, Part A: Molecular and Integrative Physiology* **133**: 95-104.

Sanders, M.E., Morelli, L. and Tompkins, T.A. 2003. Spore formers as human probiotics: *Bacillus, Sporolactobacillus and Brevibacillus*. *Comprehensive Reviews in Food Science and Food Safety* **2**: 101-110.

Santoso, U., Tanaka, K. and Ohtani, S. 1995. Effect of dried *Bacillus subtilis* culture on growth, body composition and hepatic lipogenic enzyme activity in female broiler chicks. *British Journal of Nutrition* **74**: 523-529.

SCAN. 1999. Assessment by the Scientific Committee on Animal Nutrition (SCAN) of a microorganism product: Neoferm BS-10. European Commission, Health and Consumer Protection Directorate-General. (SCAN) Scientific Committee on Animal

Nutrition. Available at: http://europa.eu.int/comm/food/fs/sc/scan/out28_en.pdf.

SCAN. 2000a. Opinion of the Scientific Committee on Animal Nutrition on the safety of use of *Bacillus* species in animal nutrition. European Commission, Health and Consumer Protection Directorate-General. (SCAN) Scientific Committee on Animal Nutrition. Available at: http://europa.eu.int/comm/food/fs/sc/scan/out41_en.pdf.

SCAN. 2000b. Report of the Scientific Committee on Animal Nutrition on product BioPlus 2B for use as a feed additive. European Commission, Health and Consumer Protection Directorate-General. (SCAN) Scientific Committee on Animal Nutrition. Available at: http://europa.eu.int/comm/food/fs/sc/scan/out49_en.pdf.

SCAN. 2001a. Guidelines for the assessment of additives in feedingstuffs. Part II: Enzymes and microorganisms. European Commission, Health and Consumer Protection Directorate-General. (SCAN) Scientific Committee on Animal Nutrition. Available at: http://europa.eu.int/comm/food/fs/sc/scan/out68_en.pdf.

SCAN. 2001b. Report of the Scientific Committee on Animal Nutrition on product Toyocerin for use as an animal feed additive. European Commission, Health and Consumer Protection Directorate-General. (SCAN) Scientific Committee on Animal Nutrition. Available at: http://europa.eu.int/comm/food/fs/sc/scan/out72_en.pdf.

SCAN. 2001c. Assessment by the Scientific Committee on Animal Nutrition on the safety of product Paciflor for use as a feed additive. European Commission, Health and Consumer Protection Directorate-General. (SCAN) Scientific Committee on Animal Nutrition. Available at: http://europa.eu.int/comm/food/fs/sc/scan/out62_en.pdf.

SCAN. 2003. Opinion of the Scientific Committee on Animal Nutrition on the criteria for assessing the safety of microorganisms resistant to antibiotics of human and veterinary importance. European Commission, Health and Consumer Protection Directorate-General. (SCAN) Scientific Committee on Animal Nutrition. Available at: http://europa.eu.int/comm/food/fs/sc/scan/out108_en.pdf.

Senesi, S. 2004. *Bacilllus* spores as probiotic products for human use. In: Ricca, E., Henriques, A.O. and Cutting, S.M., editors. *Bacterial spore formers: probiotics and emerging applications:* 131-141. Horizon Bioscience, UK.

Setlow, P. 2003. Spore germination. *Current Opinion in Microbiology* **6**: 550-556.

Shires, A., Thompson, J.R., Turner, B.V., Kennedy, P.M. and Goh, Y.K. 1987. Rate of passage of corn-canola meal and corn-soybean meal diets through the gastrointestinal tract of broiler and White Leghorn chickens. *Poultry Science* **66**: 289-298.

Spinosa, M.R., Braccini, T., Ricca, E., De Felice, M., Morelli, L., Pozzi, G. and Oggioni, M.R. 2000. On the fate of ingested *Bacillus* spores. *Research in Microbiology* **151**: 361-368.

SSC. 1999. Opinion of the Scientific Steering Committee (SSC) on antimicrobial resistance. European Commission, Health and Consumer Protection Directorate-General. (SSC) Scientific Steering Committee. Available at: http://europa.eu.int/comm/food/fs/sc/ssc/out50_en.pdf.

Stavric, S. 1992. Defined cultures and prospects. *International Journal of Food Microbiology* **15**: 245-263.

Stavric, S. and D'Aoust, J.Y. 1993. Undefined and defined bacterial preparations for the competitive exclusion of *Salmonella* in poultry – a review. *Journal of Food Protection* **56**: 173-180.

Stragier, P. and Losick, R. 1996. Molecular genetics of sporulation in *Bacillus subtilis*. *Annual Reviews in Genetics* **30**: 297-241.

Tam, N.K., Uyen, N.Q., Hong, H.A., Duc, L.H., Hoa, T.T., Serra, C.R., Henriques, A.O. and Cutting, S.M. 2006. The intestinal life cycle of *Bacillus subtilis* and close relatives. *Journal of Bacteriology* **188**: 2692-2700.

Uyen, N.Q., Hong, H.A. and Cutting, S.M. 2007. Enhanced immunisation and expression strategies using bacterial spores as heat-stable vaccine delivery vehicles. *Vaccine* **25**: 356-365.

von Wright, A. 2005. Regulating the safety of probiotics – the European approach. *Current Pharmaceutical Design* **11**: 17-23.

Wise, M.G. and Siragusa, G.R. 2007. Quantitative analysis of the intestinal bacterial community in one- to three-week-old commercially reared broiler chickens fed conventional or antibiotic-free vegetable-based diets. *Journal of Applied Microbiology* **102**: 1138-1149.

Dietary-based gut flora modulation against *Clostridium difficile* onset

Christos Gougoulias, Kieran M. Tuohy and Glenn R. Gibson

Food Microbial Sciences Unit, Department of Food Biosciences, University of Reading, Whiteknights, Reading RG6 6AP, UK. Tel. +44 118 378 4537. Fax +44 118 931 0080. E-mail c.gougoulias@reading.ac.uk

Abstract

Clostridium difficile infection is a frequent complication of antibiotic therapy in hospitalised patients, which today is attracting more attention than ever and has led to its classification as a 'superbug'. Disruption of the composition of the intestinal microflora following antibiotic treatment is an important prerequisite for overgrowth of *C. difficile* and the subsequent development of an infection. Treatment options for antibiotic-associated diarrhoea and *C. difficile*-induced colitis include administration of specific antibiotics (e.g. vancomycin), which often leads to high relapse rates. More importantly, both the rate and severity of *C. difficile*–associated diseases are increasing, with new epidemic strains of *C. difficile* often implicated. For the prevention and treatment of antibiotic-associated diarrhoea and *C. difficile* infection, several probiotic bacteria such as selected strains of lactobacilli (especially *Lactobacillus rhamnosus* GG), *Bifidobacterium longum*, and *Enterococcus faecium* and the non-pathogenic yeast *Saccharomyces boulardii* have been used. Controlled trials indicate a benefit of *S. boulardii* and *L. rhamnosus* GG as therapeutic agents when used as adjuncts to antibiotics. However, the need for more well-designed controlled trials with probiotics is explicit.

Keywords: probiotics, *Clostridium difficile*, diarrhoea

1. Introduction

Clostridium difficile is a ubiquitous anaerobic, sporulating, Gram-positive, rod-shaped bacterium that is isolated from both soil and living organisms, including humans (Smith and King 1962; Alpern and Dowell 1971; Hafiz *et al.* 1975; Hafiz and Oakley 1976; Borriello *et al.* 1983). One of the primary habitats for *C. difficile* is the gut of infants (Levett 1986). Prevalence rates of 46% have been reported in healthy infants under 12 months old, and colonisation rates of up to 55% have been reported in neonatal intensive care units (Donta and Myers 1982). Although in many infants faecal cytotoxic titres are extremely high and similar to those seen in adults with pseudomembranous colitis (PMC), infants show no obvious symptoms (Cooperstock and Zedd 1983). However, they are not resistant to the disease as a result of selective colonisation with non-toxigenic strains. In healthy adults, prevalence rates are rare at 3–5% (Viscidi *et al.* 1981), and in hospital in-patients rates are 16–35% (McFarland *et al.* 1989). The reasons for high colonisation rates with no symptoms in infants and low rates in adults are not clear.

Food Science and Technology Bulletin: Functional Foods **4** (4) 31–41
DOI: 10.1616/1476-2137.14986. Published 12 October 2007
ISSN 1476-2137

2. Importance of *C. difficile* in human infections

In 1935, Hall and O'Toole isolated *C. difficile* from the stools of healthy newborn infants and showed strains to be toxigenic. However, the bacterium was not associated with intestinal disease until the late 1970s. Antibiotic-associated diarrhoea (AAD) is defined as diarrhoea developing from a few hours after the onset of antibiotic therapy to 6–8 weeks following antibiotic discontinuation (Bartlett 1992; Hogenauer *et al.* 1998). The incidence of AAD in the literature varies from 5–25% in patients receiving antibiotics, depending on the class of antibiotic used and confounding risk factors in the patients being treated (Bartlett *et al.* 1978a; Hove *et al.* 1996; McFarland 1998). After 1950 and with the development of broad spectrum antibiotics, AAD became a recognised condition. However, it was in 1978 with the emergence of PMC occurring in patients treated with clindamycin, that *C. difficile* was identified as a serious aetiological agent (Bartlett *et al.* 1978a, 1978b; George *et al.* 1978).

3. Pathogenesis of AAD

During most courses of antibiotic therapy, alteration of the normal flora of the bowel leads to a loss of equilibrium and consequently of the resistance to colonisation, which may result in the emergence of pathogenic organisms such as *C. difficile* (Hogenauer *et al.* 1998). Disruption of the ecological equilibrium of the normal intestinal microflora may result in diarrhoea associated with alteration of fermentation processes and a reduction in short-chain fatty acids ('functional diarrhoea'). The most important bacterial aetiology of AAD is *C. difficile* (Hove *et al.* 1996).

3.1 C. difficile *and AAD*

Since the demonstration of the role of this organism in PMC (Bartlett *et al.* 1978b; George *et al.* 1978), its importance has grown significantly over the last 30 years. The levels of *C. difficile* in AAD were demonstrated to be 20–25%, with over 95% of PMC involving the microorganism (Hogenauer *et al.* 1998; McFarland *et al.* 1999; Bartlett 2002).

3.2 C. difficile *virulence factors*

The pathogenesis of PMC is mediated by two potent, heat-labile cytotoxic toxins produced by *C. difficile*: toxin A, which is an enterotoxin, and toxin B, a cytotoxin (Lyerly *et al.* 1988; Pothoulakis and Lamont 2001). Toxin A has a molecular mass of 308 kDa whereas toxin B has a molecular of 279 kDa (Barroso *et al.* 1990; Dove *et al.* 1990). The amino acid sequences of the toxins show a high level of homology (von Eichel-Streiber *et al.* 1990, 1992).

Both toxins are capable of inflicting significant damage to the human colonic epithelium, including modulating fluid secretion and inducing a necrotic inflammatory response, and they act synergistically in this (Savidge *et al.* 2003). A large range of clinical presentations have been reported, from an asymptomatic carriage to fulminant colitis, depending on the potential overgrowth of the organism. Recent work has revealed the cellular mechanism of action of the toxins (see Figure 1). Both toxins A and B have monoglycosyltransferase activity, which catalyses the incorporation of glucose into a variety of substrate proteins (Just *et al.* 1995a, 1995b, 1995c). These include the small GTP-binding proteins (Rho, Rac and Cdc42Hs) that are involved in the regulation of the actin cytoskeleton, specifically in the formation of actin stress fibres and focal adhesions. In the diseased state, the colonic epithelium is the major target of *C. difficile* toxins. They cause disruption of the barrier function by opening the tight junctions. This effect is not only caused by the breakdown of actin filaments, but also by inactivation of Rho's ability to regulate tight junction complexes. These barrier-disrupting effects of toxins A and B increase the colonic permeability, the basis of watery diarrhoea, which is a typical feature of *C. difficile* AAD (Poxton *et al.* 2001). The colitis is characterised by a massive influx of neutrophils into the colonic mucosa, and in PMC there is an acute inflammatory infiltrate with microabscesses and pseudomembranes rich in neutrophils (Souza *et al.* 1997).

3.3 Pathogenesis of alterations in the function of the intestinal flora

The normal production of lactic acid and short-chain fatty acids (acetate, butyrate, propionate) by the anaerobic flora in AAD is decreased due to diminished digestion of carbohydrates, which results in functional disturbances of the colonic mucosa (Hove *et al.* 1996; Hogenauer *et al.* 1998). Moreover, diminished or suppressed carbohydrate metabolism may result in osmotic diarrhoea ('overload mechanism') and poor absorption of short-chain fatty acids ('underload mechanism'; Clausen *et al.* 1991; Gustafsson *et al.* 1998), water and electrolytes (cations bound by anionic organic acids; Hammer *et al.* 1990). Decreased bile acid dehydroxylation, a process that is performed normally by bacteria in the colon (Takamine and Imamura 1995), has been also advocated among the metabolic disturbances resulting from antibiotic use (Hofmann 1977).

3.4 Complications

One of the most frequent complications in hospital patients with *C. difficile*-associated diarrhoea is the frequency of relapse (McFarland *et al.* 1999). This recurrent form of AAD leads to increased use of antibiotics (vancomycin), extended hospital stays and medical complications. *C. difficile*-associated disease (CDAD) causes death in 1–2% of affected patients, whereas the mortality rate increases to 6–30% when PMC is present (Miller *et al.* 2002; Aslam *et al.* 2005).

3.5 Epidemiology and risk factors

Two main types of predisposing factors have been recognised in AAD: the class of antibiotic administered, and host factors (age and underlying pathologies). Other environmental circumstances may also contribute towards the spread of PMC.

3.5.1 Antibiotic class

Nearly all antibiotics have been reported to be associated with AAD and CDAD, provided that antibiotic concentrations are high enough in the intestinal lumen to inhibit the anaerobes (Wistrom *et al.* 2001). The most commonly implicated are clindamycin, aminopenicillins (ampicillin/amoxicillin), a combination of amoxicillin and clavulanic acid, second or third generations of cephalosporins (cefuroxime, cefotaxime, ceftazidime, ceftriaxone) and, more recently, fluoroquinolones (Johnson *et al.* 1999; Wistrom

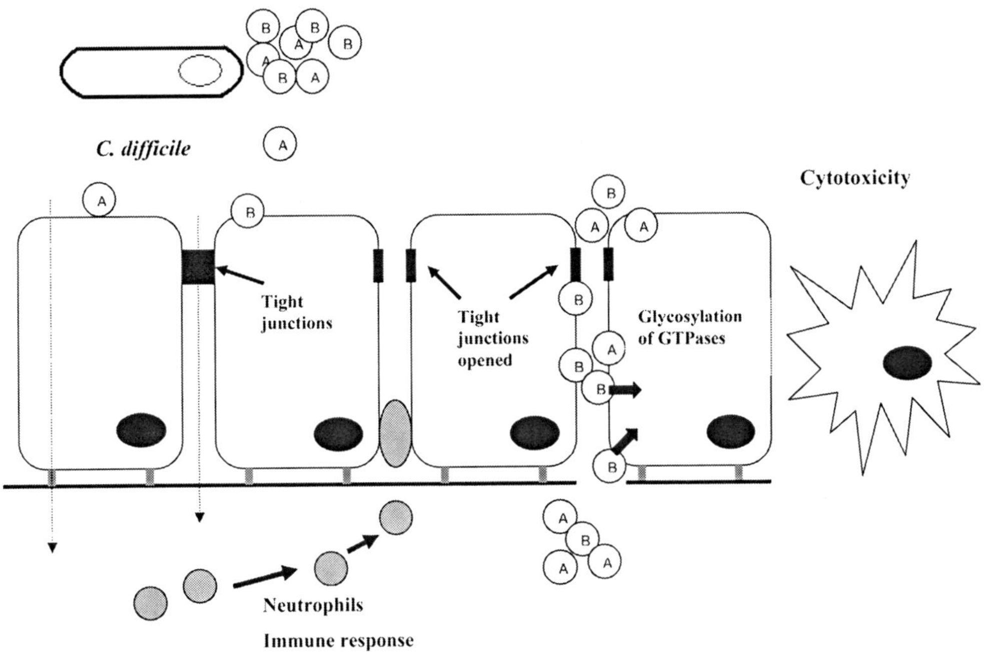

Figure 1. Mechanism of action of *Clostridium difficile* toxins A and B on the intestinal epithelium. Both toxins modify small proteins (GTP-binding proteins) that regulate the actin cytoskeleton by incorporating a glucose molecule into them. The toxins then cause disruption of the barrier function of the intestinal epithelium by opening the tight junctions and increasing colonic permeability. This is because GTPases regulate tight junction complexes. The immune response is characterised by a massive influx of neutrophils. Reproduced with permission from Ian R. Poxton, University of Edinburgh, UK.

et al. 2001; McCusker *et al.* 2003; Loo *et al.* 2005; Muto *et al.* 2005; Pepin *et al.* 2005b). In hospitals, the incidence of AAD is highest in intensive care unit (ICU) patients, with the duration of use of the inciting antibiotic therapy an additional risk factor (Wistrom *et al.* 2001). In addition, proton pump inhibitors have been implicated as a further possible risk factor (Cunningham *et al.* 2003; Dial *et al.* 2004, 2005, 2006; Yearsley *et al.* 2006).

3.5.2 Host factors

ICU patients with severe underlying pathologies, such as elderly or immunosuppressed patients, or those that have undergone surgical (transplant, gastrointestinal) procedures or that have nasogastric feed tubing, show the highest incidence of AAD and PMC (Brown *et al.* 1990). While approximately 70% of the population have serum antibodies against toxin A and/or B, the presence of basal levels of antibodies in inpatients exposed to *C. difficile* is not protective against colonisation (Kyne *et al.* 2000; Johal *et al.* 2004). Patient age is one risk factor, with a higher incidence of colonisation at the extreme ages of life (<6 and >65 years old: Brown *et al.* 1990; Ackermann *et al.* 2005).

3.5.3 Environmental factors

A hospital offers multiple opportunities for gastrointestinal, nosocomial infections, and the spread of *C. difficile* between patients has been well-documented (Wistrom *et al.* 2001). *C. difficile* is considered a nosocomial organism (Bartlett 2002) and gut surgery and gastrointestinal exploratory procedures increase the risk of AAD (Ackermann *et al.* 2005).

3.6 Treatments of AAD

3.6.1 Mild or moderate cases of AAD

Asymptomatic carriers of *C. difficile* should not be treated (Johnson *et al.* 1992). Treatments for AAD comprise conventional measures such as rehydration and discontinuation of the inciting agent, or replacement of the latter if necessary by a more appropriate antibiotic.

3.6.2 Antibiotic therapy in severe AAD: treatment of PMC

Severe cases of AAD related to *C. difficile* require a suitably adapted oral antibiotic therapy. The recommended antibacterial agents are metronidazole, which constitutes first-line therapy (Gerding 2005; Modena *et al.* 2006), and vancomycin (Teasley *et al.* 1983). Agents used less frequently include bacitracin (Young *et al.* 1985), teicoplanin and fusidic acid (Wenisch *et al.* 1996; Wullt and Odenholt 2004). However, relapses have been reported in all cases (Fernandez *et al.* 2004; Musher *et al.* 2005; Pepin *et al.* 2005a). Moreover, there is concern regarding the selection of vancomycin-resistant organisms, mainly *Enterococcus faecium*, which is a promiscuous microorganism with respect to the transfer of antibiotic resistance (ASHP 1998).

4. Changes in *C. difficile* epidemiology

Historically, low rates of severe disease and death ($\leq 3\%$) may have led to an underestimation of the importance of CDAD as an infection linked to health care (Rubin *et al.* 1995). Nevertheless, each case of CDAD has been estimated to result in more than $3600 of health care costs, and these costs exceeded $1 billion in the US in 2002 (Kyne *et al.* 2002). Both the rate and the severity of CDAD may be increasing in US health care facilities. An analysis of data from the National Nosocomial Infections Surveillance system identified an increase in rates from the late 1980s through to 2001 (Archibald *et al.* 2004).

The epidemiology of *C. difficile* is changing, as documented in many recently published articles (Pepin *et al.* 2004). Researchers from the US and Canada note increased rates of CDAD and episodes of more serious disease, both in hospitals and in the community (Pepin *et al.* 2004). The risk factors for CDAD may also be changing. As noted recently in Europe, North American researchers are also describing a new epidemic strain of *C. difficile* named BI/NAP1, which is positive for binary toxin and carries the virulence properties and antibiotic resistance patterns of the European strain (ribotybe) O27. Current BI/NAP1 strains are more resistant to fluoroquinolones than are historic isolates (McDonald *et al.* 2005). Canadian researchers also documented it as the predominant strain in a multi-institutional outbreak in the Quebec area. In a prospective study of CDAD in 1703 patients at 12 hospitals in Quebec, Loo *et al.* (2005) linked the epidemic strain to an increased incidence of *C. difficile* from 5–6 cases per 1000 admissions to about 25 per 1000 admissions. They also reported that the 30 day mortality rate attributable to infection was 6.9%. The strain was also resistant to fluoroquinolines (Loo *et al.* 2005).

Community-based cases are also changing. Recently, in the *Morbidity and Mortality Weekly Report*, four state health departments in the US reported on 33 unusual cases of community-based CDAD that had occurred between 2003 and the middle of 2005 (Anon. 2005). The cases were unusual as disease was severe and involved peripartum women and healthy individuals who had not been recently hospitalised, including eight who had no history of recent antibiotic use.

5. Rationale for living microorganisms (probiotics) in the treatment or prevention of AAD

The human normal intestinal flora is formed of up to 10^{12} bacteria per gram of intestinal content and consists of more than 1000 species (Berg 1996; Suau *et al.* 1999; Vaughan *et al.* 2000; Hughes *et al.* 2001; Blaut *et al.* 2002; Guarner and Malagelada 2003; Xu and Gordon 2003). Aerobic, facultative and anaerobic bacteria inhabit the gastrointestinal tract. The proportion of anaerobic bacteria increases from the proximal to distal regions, and 99% of inhabitants located in the large intestine are anaerobes. Despite the complexity of the gut bacterial population, its gross composition is remarkably stable and tends to be characteristic for each individual (Zoetendal *et al.* 1998, 2001; Mai *et al.* 2004; Vanhoute *et al.* 2004). All of these bacteria form a stable ecosystem together with the intestinal mucosa (Guarner and Malagelada 2003). The equilibrium of the ecosystem is affected by specific niches for individual bacterial populations, and cooperation in terms of the metabolism of various substrates and cross-feeding between bacteria (Belenguer *et al.* 2006). Therefore, one important property of this stable ecosystem is colonisation resistance, which permits the elimination of exogenous microorganisms. However, any antibiotic therapy may alter the normal equilibrium of the intestinal microflora, hence encouraging the potential for pathogenic organisms to emerge and cause abnormal growth of *C. difficile* (Johnson *et al.* 1999; Wistrom *et al.* 2001; Pepin *et al.* 2005b). The hypothesis that living microorganisms could be administered for treatment or prevention of AAD has been supported by the administration of several 'physiological' non-pathogenic organisms (Table 1). These have been designated 'probiotics' (Cremonini *et al.* 2002b; Szajewska *et al.* 2006) or 'biotherapeutic agents' (Elmer *et al.* 1996; Roffe 1996), and defined as "living microbial supplements that exert a beneficial effect on the host by improving the intestinal ecosystem". Most of these probiotics are bacteria or yeasts used clinically in lyophilised form, and are commercially available.

5.1 Probiotics in the prevention and treatment of AAD

Both bacteria and yeast have been used in the treatment and prevention of AAD, as well as the prevention of associated diarrhoea and reduced relapse (Table 1).

Lactobacillus rhamnosus GG was administered in the form of a yoghurt for 1 week prior to administration of

Table 1. Examples of clinical trials using living microorganisms in the treatment and prevention of antibiotic-associated diarrhoea (AAD)

Biotherapeutic agent	Indications	Type of study	No. patients	Dose (per day) and duration of treatment	Results	Reference
Lactobacillus rhamnosus GG	Treatment of AAD	RPC	16	–	Shortened duration of diarrhoea (8 days)	(Siitonen *et al.* 1990)
Lactobacillus acidophilus + *Lactobacillus bulgaricus*	Prevention of AAD	DBPC	38	20.4×10^8 CFU/day, for 10 days (min 5 days)	No significant prevention in the treated group	(Tankanow *et al.* 1990)
Bifidobacterium longum	Prevention of AAD	RPC	10	3 yoghurts/day, for 3 days	No significant increase in weight and frequency of stools in treated group vs. placebo	(Colombel *et al.* 1987)
L. rhamnosus GG	Prevention of AAD	DBPC	188	1×10^{10} CFU to 2×10^{10} CFU/day, for 10 days	17% AAD vs. 48% in placebo group	(Young *et al.* 1998)
Saccharomyces boulardii	Prevention of AAD	DBPC	388	4 capsules/day, variable duration of treatment (min 5 days)	4.5% AAD vs. 17.5% in placebo group	(Adam *et al.* 1977)
S. boulardii	Prevention of AAD	DBPC	180	1 g/day, variable duration of treatment	9.5% AAD vs. 14.6% in placebo group	(Surawicz *et al.* 1995)
S. boulardii	Prevention of AAD	DBPC	72	113 mg twice a day, for 49 days	No significant prevention in the treated group	(Lewis *et al.* 1998)
S. boulardii	Prevention of AAD	DBPC	193	1 g/day, for 49 days	7.2% AAD vs. 14.6% in placebo group	(McFarland *et al.* 1995)
S. boulardii	Prevention of AAD	DBPC	269	500 mg/day, for the duration of antibiotic treatment (experimental group 7.8±1 day; control group 8.1±1 day)	3.4% AAD vs. 17.3% in placebo group	(Kotowska *et al.* 2005)
S. boulardii	Prevention of AAD	DBPC	151	500 mg/day, for the duration of antibiotic treatment	1.4% AAD vs. 9% in placebo group ($p < 0.05$)	(Can *et al.* 2006)

(*Continued*)

Table 1. (*Continued*)

Biotherapeutic agent	Indications	Type of study	No. patients	Dose (per day)and duration of treatment	Results	Reference
L. acidophilus + *Bifidobacterium bifidum*	Prevention of AAD	DBPC	150	2×10^{10} CFU each strain/day, for 20 days	2.9% toxins vs. 7.3% in placebo group	(Plummer *et al.* 2004)
Enterococcus faecium SF68	Prevention of AAD	DBPC	45	7.5×10^{7} CFU/day, for 7 days	9% AAD vs. 27% in placebo group	(Wunderlich *et al.* 1989)
L. rhamnosus GG	Prevention of relapses of PMC	Open	5	–	No relapses in four patients, metronidazole in one patient	(Gorbach *et al.* 1987)
S. boulardii	Prevention of relapses of PMC	DBPC	124	1 g/day, for 28 days	Further relapses in 26% of treated patients vs. 45% in placebo group	(McFarland *et al.* 1994)
S. boulardii	Prevention of relapses of CDAD	Open	19	–	Diminished number of relapses	(Buts *et al.* 1993)
Lactobacillus plantarum 299v	Prevention of relapses of CDAD	DBPC	20	5×10^{10} CFU/day, for 38 days	Recurrence rate 36% vs. 67% in placebo group	(Wullt *et al.* 2003)
L. plantarum 299v	Reduction of the negative effects of an antibiotic on colonic fermentation	DBPC	19	–	Significant decrease in total SCFA in the placebo group ($p = 0.028$) but not in the *Lactobacillus* group	(Wullt *et al.* 2007)
S. boulardii	Prevention of relapses of CDAD	DBPC	209	1 g/day, for 28 days	Recurrence rate 17% vs. 50% in placebo group with high-dose vancomycin. No difference in recurrence rate in treated group vs. placebo group with low-dose vancomycin or metronidazole	(Surawicz *et al.* 2000)

CDAD: *Clostridium difficile*-associated diarrhoea; DBPC: double blind placebo controlled; PMC: pseudomembranous colitis; RPC: randomized placebo controlled; SCFA: short-chain fatty acids; -: data not available erythromycin (400 mg three times/day). This resulted in a diminished duration of diarrhoea in 16 patients compared to a placebo group (2 vs. 8 days; $p < 0.05$) and *C. difficile* toxin was negative in all 16 patients at the end of the treatment period (Siitonen *et al.* 1990). In 38 paediatric patients with amoxicillin-induced diarrhoea the administration of a mixture of *Lactobacillus acidophilus* and *Lactobacillus bulgaricus*, four times a day for 10 days together with the antibiotic therapy, had no preventative effect on diarrhoea (Tankanow *et al.* 1990). Stool samples were not tested for *C. difficile* (culture and/or toxin). However, in a randomised placebo-controlled cross-over study in 10 healthy volunteers receiving erythromycin (3 days, 1 g twice/day), oral ingestion of yoghurt containing *Bifidobacterium longum* significantly diminished the number of stools (double in the placebo group but unchanged in the yoghurt group) and degree of abdominal pain compared to volunteers receiving placebo-yoghurt (Colombel *et al.* 1987). Stool samples were not tested for *C. difficile* (culture and/or toxin). In 188 children (6 months to 10 years old) treated with antibiotics, administration of *Lactobacillus* GG (1–2 capsules a day, 10^{10} CFU per capsule) resulted in a significant difference in the occurrence of diarrhoea (16 out of 93 children) compared to the placebo group (46 out of 95 children, $p < 0.0001$; Young *et al.* 1998). In a recent double-blind, randomised, placebo-controlled trial, 150 elderly patients received standard antibiotic treatment plus 2×10^{10} CFU *L. acidophilus* and *Bifidobacterium bifidum*/capsule per day for 20 days. On the basis of the development of diarrhoea, the incidence of samples positive for *C. difficile*-associated toxins was 2.9% in the probiotic group vs. 7.25% in the placebo group. When samples from all patients were tested, 46% of probiotic patients were toxin-positive compared with 78% of the placebo group (Plummer *et al.* 2004).

E. faecium SF68 has shown only modest efficacy in the prevention of AAD in two controlled clinical trials. Forty-five patients being treated with antibiotics were given, concurrently, one capsule twice daily of either *E. faecium* SF68 (7.5×10^7 CFU) or placebo for 7 days. The rate of AAD was 9% with SF68 compared with a placebo rate of 27% (Wunderlich *et al.* 1980). In the second study (not double-blind), 200 patients received antituberculic antibiotics. Patients administered orally with *E. faecium* SF68 showed a lower rate (5%) of AAD compared to the placebo group (18%; Borgia *et al.* 1982).

In 388 ambulatory patients with upper respiratory tract infections receiving tetracycline or β-lactams for longer than 5 days, *Saccharomyces boulardii* reduced the prevalence of AAD to 4.5% compared to 17.5% in the placebo group ($p < 0.001$; Adam *et al.* 1977). Stool samples were not tested for *C. difficile* (culture and/or toxin). In a double-blind, placebo-controlled study with 180 patients receiving any antibiotic except vancomycin or metronidazole, the occurrence of AAD was limited to 9.5% in patients treated with *S. boulardii* compared to 22% in the placebo group (Surawicz *et al.* 1995). Stool samples were also tested for *C. difficile* (culture and toxin). Similarly, in 193 patients receiving β-lactam antibiotics, the administration of *S. boulardii* (1 g daily for 2 weeks) in addition to antibiotics, resulted in a lower incidence of AAD (7.2% of patients receiving *S. boulardii* compared to 14.6% in the placebo group, $p = 0.02$; McFarland *et al.* 1995). Simultaneous administration is possible because of its resistance to most antibiotics (Bergogne-Berezin 1995). However, Lewis *et al.* (1998) reported a lack of a therapeutic effect with *S. boulardii* (113 mg twice daily) in the prevention of AAD in a placebo-controlled study with 72 elderly patients. The lack of effect was attributed to the low dose used. However, in a recent double-blind, randomised placebo-controlled trial of 269 children (aged 6 months to 14 years) with otitis and/or respiratory tract infections, standard antibiotic treatment together with 250 mg of *S. boulardii* orally twice daily resulted in a lower prevalence of diarrhoea than in those receiving antibiotic plus placebo (8 vs. 23%). In a separate study, the occurrence of AAD was also limited to 3.4% in patients treated with *S. boulardii* compared to 17.3% in the placebo group (Kotowska *et al.* 2005). One hundred and fifty-one patients hospitalised at the Gulhane Military Medical Academy's Department of Infectious Diseases and Clinical Microbiology were given *S. boulardii* twice daily during the course of antibiotic therapy. Patients did not require intensive care therapy and application was initiated as late as 48 h after antibiotic therapy. The rate of development of AAD was found to be 9% (7/78) in the placebo group was and 1.4% (1/73) in the study group ($p < 0.05$). Stool samples from the patients with AAD were assayed for *C. difficile* toxin A as late as 7 days after the collection of samples. The assay for *C. difficile* toxin A yielded positive results in two stool samples from the seven patients with AAD in the placebo group and a negative result was obtained for the one patient who developed AAD in the study group. The results implied that prophylactic use of *S. boulardii* in hospitalised patients resulted in a reduced incidence of AAD, with no serious side effects seen (Can *et al.* 2006).

5.2 *Probiotics in the prevention and treatment of CDAD*

Early uncontrolled trials of *Lactobacillus rhamnosus* GG suggested a possible efficacy in recurrent CDAD. In four different studies, 8 out of 11 adults (Gorbach *et al.* 1987), two out of four children (Biller *et al.* 1995), five out of nine adults (Bennett *et al.* 1990) and 27 out of 32 adults (Bennett *et al.* 1996) demonstrated beneficial effects.

A single randomised, controlled trial using *L. rhamnosus* GG with either metronidazole or vancomycin has been reported (Pochapin 2000). The recurrence rate was similar for the probiotic (36.4%) and placebo groups (35.7%). A small study of 20 patients with recurrent CDAD, metronidazole and either *Lactobacillus plantarum* 299v (5 × 10^{10} CFU/day) or placebo were given for 38 days (Wullt *et al.* 2003). Although the recurrence rate was lower in the *L. plantarum* group (36.4%) compared to the placebo (66.7%), the difference was not significant. In 19 children with persistent diarrhoea associated with the presence of *C. difficile* in stools, *S. boulardii* administered as a monotherapy resulted in a significant improvement in symptoms in 18 patients after 8 days of treatment (Buts *et al.* 1993). In a double-blind, randomised, controlled study, *S. boulardii* versus placebo was given to 124 patients with *C. difficile* associated-diarrhoea, with a view of preventing relapses. The number of relapses was significantly reduced to 26% compared to the placebo group at 45% ($p = 0.05$). The efficacy of *S. boulardii* in preventing relapses has been shown more clearly in patients who had a previous relapse of AAD. In these patients, relapses occurred in 35 vs. 65% ($p = 0.04$) in the placebo control (McFarland *et al.* 1994). In an effort to further refine a standard regimen, the same group tested patients receiving a standard antibiotic for 10 days and then added either *S. boulardii* (1 g/day for 28 days) or placebo. A significant decrease in the recurrence was observed only in patients treated with high-dose vancomycin (2 g/day) and *S. boulardii* (17%), compared to those who received high-dose vancomycin and placebo (50%, $p = 0.05$). However, *S. boulardii* treatment had no impact on the recurrence rates in patients treated with a low dose of vancomycin or metronidazole (Surawicz *et al.* 2000).

A recent clinical trial demonstrated that administration of *L. plantarum* 299v reduced the negative effects of metronidazole on colonic fermentation (Wullt *et al.* 2007). The authors suggested that intake of *L. plantarum* 299v affected concentrations of faecal organic acids during and after metronidazole treatment in 19 patients with recurrent *C. difficile*-associated diarrhoea. Following the intake of metronidazole, a significant decrease in total short-chain fatty acids was seen in the placebo group (from 77.1 to 45.5 μmol/g, $p = 0.028$), but no effect was seen in the *Lactobacillus* group (79.8 to 60.4 μmol/g). In addition, a statistically significant difference between treatment groups was noted for butyrate (from 5.6 to 1.2 μmol/g in the placebo group vs. 7.6 to 5.6 μmol/g in the *Lactobacillus* group, $p = 0.047$). At the end of the study and after cessation of placebo or *Lactobacillus* treatment, total short-chain fatty acids rose to the same levels as those recorded prior to the start of antibiotic treatment in the placebo group. Therefore it is possible that intake of this probiotic strain may provide an additional benefit for patients with recurrent *C. difficile*-associated diarrhoea.

6. Conclusions

AAD and *C. difficile* are common clinical problems. The concept of replacing disease-inducing pathogenic organisms with non-pathogenic ones appears useful. The use of probiotics in the control of gastrointestinal disorders offers an alternative approach/adjunct to conventional antibiotic therapies. Several live microorganisms have been used in the treatment and prevention of AAD and CDAD. They include *L. rhamnosus* GG, *B. bifidum* or yeasts such as *S. boulardii*. The results discussed here show an overall reduction in the risk of AAD during probiotic administration. The most reproducible results so far have been achieved with *S. boulardii* and *L. rhamnosus* GG predominantly, both in terms of reducing the incidence of AAD and the number of relapses in recurrent *C. difficile*-associated diarrhoea. However, the number of clinical trials with 'biotherapeutic' agents is limited, and one of the main issues is inconsistency; many studies on AAD have not tested for *C. difficile* (culture and/or toxins) before or after treatment, the definition of diarrhoea within the studies varies, the duration and the dose of treatment varies as does the follow-up period, and the number of patients and the design in many (early) studies is considered inadequate. Some theoretical considerations also arise regarding the possibility of side-effects with 'biotherapeutic' agents such as bacteraemia, especially in immunosuppressed patients. These concerns have, however, not been confirmed by clinical experience. Therefore, further well-designed, controlled clinical trials with standard definitions and fixed parameters need to be undertaken, using various probiotic preparations for the treatment of confirmed *C. difficile* diarrhoea. In conclusion, it would appear that there is some merit in the probiotic approach to address *C. difficile*-induced problems. Certainly, there is a need to address the ubiquity and consequences of this pathogen.

7. References

Ackermann, G., Thomalla, S., Ackermann, F., Schaumann, R., Rodloff, A.C. and Ruf, B.R. 2005. Prevalence and characteristics of bacteria and host factors in an outbreak situation of antibiotic-associated diarrhoea. *Journal of Medical Microbiology* **54**: 149-153.

Adam, J., Barret, A. and Barret-Bellet, L. 1977. Essais cliniques controles en double insu de l'Ultra levure lyophilisee. Etude multi-centrique par 25 medecins de 338 cas. *Gazette Medicale de France* **84**: 2072-2078.

Alpern, R.J. and Dowell, V.R. Jr. 1971. Nonhistotoxic clostridial bacteremia. *American Journal of Clinical Pathology* **55**: 717-722.

Anon. 2005. Severe CDAD in populations previously at low risk-four states, 2005. *MMWR Morbidity and Mortality Weekly Reports* **54**: 1201-1205.

Archibald, L.K., Banerjee, S.N. and Jarvis, W.R. 2004. Secular trends in hospital-acquired *Clostridium difficile* disease in the United States, 1987-2001. *Journal of Infectious Diseases* **189**: 1585-1589.

ASHP. 1998. ASHP therapeutic position statement on the preferential use of metronidazole for the treatment of *Clostridium difficile*-associated disease. *American Journal of Health System-Pharmacy* **55**: 1407-1411.

Aslam, S., Hamill, R.J. and Musher, D.M. 2005. Treatment of *Clostridium difficile*-associated disease: old therapies and new strategies. *Lancet Infectious Diseases* **5**: 549-557.

Barroso, L.A., Wang, S.Z., Phelps, C.J., Johnson, J.L. and Wilkins, T.D. 1990. Nucleotide sequence of *Clostridium difficile* toxin B gene. *Nucleic Acids Research* **18**: 4004.

Bartlett, J.G., Chang, T.W., Gurwith, M., Gorbach, S.L. and Onderdonk, A.B. 1978a. Antibiotic-associated pseudomembranous colitis due to toxin-producing clostridia. *New England Journal of Medicine* **298**: 531-534.

Bartlett, J.G., Moon, N., Chang, T.W., Taylor, N. and Onderdonk, A. B. 1978b. Role of *Clostridium difficile* in antibiotic-associated pseudomembranous colitis. *Gastroenterology* **75**: 778-782.

Bartlett, J.G. 1992. Antibiotic-associated diarrhea. *Clinical Infectious Diseases* **15**: 573-581.

Bartlett, J.G. 2002. Clinical practice. Antibiotic-associated diarrhea. *New England Journal of Medicine* **346**: 334-339.

Belenguer, A., Duncan, S.H., Calder, A.G., Holtrop, G., Louis, P., Lobley, G.E. and Flint, H.J. 2006. Two routes of metabolic cross-feeding between *Bifidobacterium adolescentis* and butyrate-producing anaerobes from the human gut. *Applied and Environmental Microbiology* **72**: 3593-3599.

Bennett, R.G., Laughon, B. and Lindsay, J. 1990. *Lactobacillus* GG treatment of *Clostridium difficile* infection in nursing home patients. *Third International Conference on Nosocomial Infection, Atlanta, Georgia, USA.*

Bennett, R.G., Gorbach, S.L., Goldin, R. and Chang, T.W. 1996. Treatment of relapsing *Clostridium difficile* diarrhea with *Lactobacillus* GG. *Nutrition Today* **31**: S35-S38.

Berg, R.D. 1996. The indigenous gastrointestinal microflora. *Trends in Microbiology* **4**: 430-435.

Bergogne-Berezin, E. 1995. Ecologic impact of antibiotherapy. Role of substitution microorganisms in the control of antibiotic-related diarrhea and colitis. *Presse Medicale* **24**: 145-156.

Biller, J.A., Katz, A.J., Flores, A.F., Buie, T.M. and Gorbach, S.L. 1995. Treatment of recurrent *Clostridium difficile* colitis with *Lactobacillus* GG. *Journal of Pediatric Gastroenterology and Nutrition* **21**: 224-226.

Blaut, M., Collins, M.D., Welling, G.W., Dore, J., van Loo, J. and de Vos, W.M. 2002. Molecular biological methods for studying the gut microbiota: the EU human gut flora project. *British Journal of Nutrition* **87**:(Suppl. 2) S203-S211.

Borgia, M., Sepe, N., Brancato, V. and Borgia, B.A. 1982. A controlled clinical study on *Streptococcus faecium* preparation for the prevention of side reactions during long-term antibiotic treatments. *Current Therapeutic Research* **31**: 266-271.

Borriello, S.P., Honour, P., Turner, T. and Barclay, F. 1983. Household pets as a potential reservoir for *Clostridium difficile* infection. *Journal of Clinical Pathology* **36**: 84-87.

Brown, E., Talbot, G.H., Axelrod, P., Provencher, M. and Hoegg, C. 1990. Risk factors for *Clostridium difficile* toxin-associated diarrhea. *Infection Control and Hospital Epidemiology* **11**: 283-290.

Buts, J.P., Corthier, G. and Delmee, M. 1993. *Saccharomyces boulardii* for *Clostridium difficile*-associated enteropathies in infants. *Journal of Pediatric Gastroenterology and Nutrition* **16**: 419-425.

Can, M., Beşirbellioglu, B.A., Avci, I.Y., Beker, C.M. and Pahsa, A. 2006. Prophylactic *Saccharomyces boulardii* in the prevention of antibiotic-associated diarrhea: a prospective study. *Medical Science Monitor* **12**(2). 19-22.

Clausen, M.R., Bonnen, H. and Mortensen, P.B. 1991. Colonic fermentation of dietary fibre to short chain fatty acids in patients with adenomatous polyps and colonic cancer. *Gut* **32**: 923-928.

Colombel, J.F., Cortot, A., Neut, C. and Romond, C. 1987. Yoghurt with *Bifidobacterium longum* reduces erythromycin-induced gastrointestinal effects. *Lancet* **2**: 43.

Cooperstock, M.S. and Zedd, A.J. 1983. *Intestinal flora of infants*. Academic Press Inc., New York, USA.

Cremonini, F., Di Caro, S., Santarelli, L. *et al.* 2002. Probiotics in antibiotic-associated diarrhoea. *Digestive and Liver Disease* **34**:(Suppl. 2) S78-S80.

Cunningham, R., Dale, B., Undy, B. and Gaunt, N. 2003. Proton pump inhibitors as a risk factor for *Clostridium difficile* diarrhoea. *Journal of Hospital Infections* **54**: 243-245.

Dial, S., Alrasadi, K., Manoukian, C., Huang, A. and Menzies, D. 2004. Risk of *Clostridium difficile* diarrhea among hospital inpatients prescribed proton pump inhibitors: cohort and case-control studies. *Canadian Medical Association Journal* **171**: 33-38.

Dial, S., Delaney, J.A., Barkun, A.N. and Suissa, S. 2005. Use of gastric acid-suppressive agents and the risk of community-acquired *Clostridium difficile*-associated disease. *Journal of the American Medical Association* **294**: 2989-2995.

Dial, S., Delaney, J.A., Schneider, V. and Suissa, S. 2006. Proton pump inhibitor use and risk of community-acquired *Clostridium difficile*-associated disease defined by prescription for oral vancomycin therapy. *Canadian Medical Association Journal* **175**: 745-748.

Donta, S.T. and Myers, M.G. 1982. *Clostridium difficile* toxin in asymptomatic neonates. *Journal of Pediatrics* **100**: 431-434.

Dove, C.H., Wang, S.Z., Price, S.B. *et al.* 1990. Molecular characterization of the *Clostridium difficile* toxin A gene. *Infection and Immunity* **58**: 480-488.

Elmer, G.W., Surawicz, C.M. and McFarland, L.V. 1996. Biotherapeutic agents. A neglected modality for the treatment and prevention of selected intestinal and vaginal infections. *Journal of the American Medical Association* **275**: 870-876.

Fernandez, A., Anand, G. and Friedenberg, F. 2004. Factors associated with failure of metronidazole in *Clostridium difficile*-associated disease. *Journal of Clinical Gastroenterology* **38**: 414-418.

George, R.H., Symonds, J.M., Dimock, F. *et al.* 1978. Identification of *Clostridium difficile* as a cause of pseudomembranous colitis. *British Medical Journal* **1**: 695.

Gerding, D.N. 2005. Metronidazole for *Clostridium difficile*-associated disease: is it okay for Mom? *Clinical Infectious Diseases* **40**: 1598-1600.

Gorbach, S.L., Chang, T.W. and Goldin, B. 1987. Successful treatment of relapsing *Clostridium difficile* colitis with *Lactobacillus* GG. *Lancet* **2**: 1519.

Guarner, F. and Malagelada, J.R. 2003. Gut flora in health and disease. *Lancet* **361**: 512-519.

Gustafsson, A., Lund-Tonnesen, S., Berstad, A., Midtvedt, T. and Norin, E. 1998. Faecal short-chain fatty acids in patients with antibiotic-associated diarrhoea, before and after faecal enema treatment. *Scandinavian Journal of Gastroenterology* **33**: 721-727.

Hafiz, S., McEntegart, M.G., Morton, R.S. and Waitkins, S.A. 1975. *Clostridium difficile* in the urogenital tract of males and females. *Lancet* **1**: 420-421.

Hafiz, S. and Oakley, C.L. 1976. *Clostridium difficile*: isolation and characteristics. *Journal of Medical Microbiology* **9**: 129-136.

Hall, I. C. and O'Toole, E. 1935. Intestinal flora in new-born infants. *American Journal of Diseases of Children* **49**: 390-402.

Hammer, H.F., Fine, K.D., Santa Ana, C.A., Porter, J.L., Schiller, L.R. and Fordtran, J.S. 1990. Carbohydrate malabsorption. Its measurement and its contribution to diarrhea. *Journal of Clinical Investigation* **86**: 1936-1944.

Hofmann, A.F. 1977. Bile acids, diarrhea, and antibiotics: data, speculation, and a unifying hypothesis. *Journal of Infectious Diseases* **135**: S126-S132.

Hogenauer, C., Hammer, H.F., Krejs, G.J. and Reisinger, E.C. 1998. Mechanisms and management of antibiotic-associated diarrhea. *Clinical Infectious Diseases* **27**: 702-710.

Hove, H., Tvede, M. and Mortensen, P.B. 1996. Antibiotic-associated diarrhoea, *Clostridium difficile*, and short-chain fatty acids. *Scandinavian Journal of Gastroenterology* **31**: 688-693.

Hughes, J.B., Hellmann, J.J., Ricketts, T.H. and Bohannan, B.J. 2001. Counting the uncountable: statistical approaches to estimating microbial diversity. *Applied and Environmental Microbiology* **67**: 4399-4406.

Johal, S.S., Lambert, C.P., Hammond, J., James, P.D., Borriello, S.P. and Mahida, Y.R. 2004. Colonic IgA producing cells and macrophages are reduced in recurrent and non-recurrent *Clostridium difficile* associated diarrhoea. *Journal of Clinical Pathology* **57**: 973-979.

Johnson, S., Homann, S.R., Bettin, K.M. *et al.* 1992. Treatment of asymptomatic *Clostridium difficile* carriers (fecal excretors) with vancomycin or metronidazole. A randomized, placebo-controlled trial. *Annals of Internal Medicine* **117**: 297-302.

Johnson, S., Samore, M.H., Farrow, K.A. *et al.* 1999. Epidemics of diarrhea caused by a clindamycin-resistant strain of *Clostridium difficile* in four hospitals. *New England Journal of Medicine* **341**: 1645-1651.

Just, I., Selzer, J., von Eichel-Streiber, C. and Aktories, K. 1995a. The low molecular mass GTP-binding protein Rho is affected by toxin A from *Clostridium difficile*. *Journal of Clinical Investigation* **95**: 1026-1031.

Just, I., Selzer, J., Wilm, M., von Eichel-Streiber, C., Mann, M. and Aktories, K. 1995b. Glucosylation of Rho proteins by *Clostridium difficile* toxin B. *Nature* **375**: 500-503.

Just, I., Wilm, M., Selzer, J., Rex, G. *et al.* 1995c. The enterotoxin from *Clostridium difficile* (ToxA) monoglucosylates the Rho proteins. *Journal of Biological Chemistry* **270**: 13932-13936.

Kotowska, M., Albrecht, P. and Szajewska, H. 2005. *Saccharomyces boulardii* in the prevention of antibiotic-associated diarrhoea in children: a randomized double-blind placebo-controlled trial. *Alimentary Pharmacology and Therapeutics* **21**: 583-590.

Kyne, L., Warny, M., Qamar, A. and Kelly, C.P. 2000. Asymptomatic carriage of *Clostridium difficile* and serum levels of IgG antibody against toxin A. *New England Journal of Medicine* **342**: 390-397.

Kyne, L., Hamel, M.B., Polavaram, R. and Kelly, C.P. 2002. Health care costs and mortality associated with nosocomial diarrhea due to *Clostridium difficile*. *Clinical Infectious Diseases* **34**: 346-353.

Levett, P.N. 1986. *Clostridium difficile* in habitats other than the human gastro-intestinal tract. *Journal of Infection* **12**: 253-263.

Lewis, S.J., Potts, L.F. and Barry, R.E. 1998. The lack of therapeutic effect of *Saccharomyces boulardii* in the prevention of antibiotic-related diarrhoea in elderly patients. *Journal of Infection* **36**: 171-174.

Loo, V.G., Poirier, L., Miller, M.A. *et al.* 2005. A predominantly clonal multi-institutional outbreak of *Clostridium difficile*-associated diarrhea with high morbidity and mortality. *New England Journal of Medicine* **353**: 2442-2449.

Lyerly, D.M., Krivan, H.C. and Wilkins, T.D. 1988. *Clostridium difficile*: its disease and toxins. *Clinical Microbiology Reviews* **1**: 1-18.

Mai, V., Katki, H.A., Harmsen, H., Gallaher, D., Schatzkin, A. *et al.* 2004. Effects of a controlled diet and black tea drinking on the fecal microbiota composition and the fecal bile acid profile of human volunteers in a double-blinded randomised feeding study. *Journal of Nutrition* **134**: 473-478.

McCusker, M.E., Harris, A.D., Perencevich, E. and Roghmann, M.C. 2003. Fluoroquinolone use and *Clostridium difficile*-associated diarrhea. *Emerging Infectious Diseases* **9**: 730-733.

McDonald, L.C., Killgore, G.E., Thompson, A. *et al.* 2005. An epidemic, toxin gene-variant strain of *Clostridium difficile*. *New England Journal of Medicine* **353**: 2433-2441.

McFarland, L.V., Mulligan, M.E., Kwok, R.Y. and Stamm, W.E. 1989. Nosocomial acquisition of *Clostridium difficile* infection. *New England Journal of Medicine* **320**: 204-210.

McFarland, L.V., Surawicz, C.M., Greenberg, R.N. *et al.* 1994. A randomized placebo-controlled trial of *Saccharomyces boulardii* in combination with standard antibiotics for *Clostridium difficile* disease. *Journal of the American Medical Association* **271**: 1913-1918.

McFarland, L.V., Surawicz, C.M., Greenberg, R.N. *et al.* 1995. Prevention of beta-lactam-associated diarrhea by *Saccharomyces boulardii* compared with placebo. *American Journal of Gastroenterology* **90**: 439-448.

McFarland, L.V. 1998. Epidemiology, risk factors and treatments for antibiotic-associated diarrhea. *Digestive Diseases* **16**: 292-307.

McFarland, L.V., Surawicz, C.M., Rubin, M., Fekety, R., Elmer, G.W. and Greenberg, R. N. 1999. Recurrent *Clostridium difficile* disease: epidemiology and clinical characteristics. *Infection Control and Hospital Epidemiology* **20**: 43-50.

Miller, M.A., Hyland, M., Ofner-Agostini, M., Gourdeau, M. and Ishak, M. 2002. Morbidity, mortality, and healthcare burden of nosocomial *Clostridium difficile*-associated diarrhea in Canadian hospitals. *Infection Control and Hospital Epidemiology* **23**: 137-140.

Modena, S., Gollamudi, S. and Friedenberg, F. 2006. Continuation of antibiotics is associated with failure of metronidazole for *Clostridium difficile*-associated diarrhea. *Journal of Clinical Gastroenterology* **40**: 49-54.

Musher, D.M., Aslam, S., Logan, N. *et al.* 2005. Relatively poor outcome after treatment of *Clostridium difficile* colitis with metronidazole. *Clinical Infectious Diseases* **40**: 1586-1590.

Muto, C.A., Pokrywka, M., Shutt, K. *et al.* 2005. A large outbreak of *Clostridium difficile*-associated disease with an unexpected proportion of deaths and colectomies at a teaching hospital following increased fluoroquinolone use. *Infection Control and Hospital Epidemiology* **26**: 273-280.

Pepin, J., Valiquette, L., Alary, M.E. *et al.* 2004. *Clostridium difficile*-associated diarrhea in a region of Quebec from 1991 to 2003: a changing pattern of disease severity. *Canadian Medical Association Journal* **171**: 466-472.

Pepin, J., Alary, M.E., Valiquette, L. *et al.* 2005a. Increasing risk of relapse after treatment of *Clostridium difficile* colitis in Quebec, Canada. *Clinical Infectious Diseases* **40**: 1591-1597.

Pepin, J., Saheb, N., Coulombe, M.A. *et al.* 2005b. Emergence of fluoroquinolones as the predominant risk factor for *Clostridium difficile*-associated diarrhea: a cohort study during an epidemic in Quebec. *Clinical Infectious Diseases* **41**: 1254-1260.

Plummer, S., Weaver, M.A., Harris, J.C., Dee, P. and Hunter, J. 2004. *Clostridium difficile* pilot study: effects of probiotic supplementation on the incidence of *C. difficile* diarrhoea. *International Microbiology* **7**: 59-62.

Pochapin, M. 2000. The effect of probiotics on *Clostridium difficile* diarrhea. *American Journal of Gastroenterology* **95**: S11-S13.

Pothoulakis, C. and Lamont, J.T. 2001. Microbes and microbial toxins: paradigms for microbial-mucosal interactions II. The integrated response of the intestine to *Clostridium difficile* toxins. *American Journal of Physiology-Gastrointestinal and Liver Physiology* **280**: G178-G183.

Poxton, I.R., McCoubrey, J. and Blair, G. 2001. The pathogenicity of *Clostridium difficile*. *Clinical Microbiology and Infection* **7**: 421-427.

Roffe, C. 1996. Biotherapy for antibiotic-associated and other diarrhoeas. *Journal of Infection* **32**: 1-10.

Rubin, M.S., Bodenstein, L.E. and Kent, K.C. 1995. Severe *Clostridium difficile* colitis. *Diseases of the Colon and Rectum* **38**: 350-354.

Savidge, T.C., Pan, W.H., Newman, P., O'Brien, M., Anton, P.M. and Pothoulakis, C. 2003. *Clostridium difficile* toxin B is an inflammatory enterotoxin in human intestine. *Gastroenterology* **125**: 413-420.

Siitonen, S., Vapaatalo, H., Salminen, S., Gordin, A., Saxelin, M., Wikberg, R. and Kirkkola, A.L. 1990. Effect of *Lactobacillus* GG yoghurt in prevention of antibiotic associated diarrhoea. *Annals of Medicine* **22**: 57-59.

Smith, L.D. and King, E.O. 1962. Occurrence of *Clostridium difficile* in infections of man. *Journal of Bacteriology* **84**: 65-67.

Souza, M.H., Melo-Filho, A.A., Rocha, M.F. *et al.* 1997. The involvement of macrophage-derived tumour necrosis factor and lipoxygenase products on the neutrophil recruitment induced by *Clostridium difficile* toxin B. *Immunology* **91**: 281-288.

Suau, A., Bonnet, R., Sutren, M. *et al.* 1999. Direct analysis of genes encoding 16S rRNA from complex communities reveals many novel molecular species within the human gut. *Applied and Environmental Microbiology* **65**: 4799-4807.

Surawicz, C.M., Elmer, G.W., Speelman, P., McFarland, L.V., Chinn, J. and Van Belle, G. 1995. Prevention of antibiotic-associated diarrhea by *Saccharomyces boulardii*: a prospective study. *Gastroenterology* **96**4. 981988.

Surawicz, C.M., McFarland, L.V., Greenberg, R.N. *et al.* 2000. The search for a better treatment for recurrent *Clostridium difficile* disease: use of high-dose vancomycin combined with *Saccharomyces boulardii*. *Clinical Infectious Diseases* **31**: 1012-1017.

Szajewska, H., Ruszczynski, M. and Radzikowski, A. 2006. Probiotics in the prevention of antibiotic-associated diarrhea in children: a meta-analysis of randomized controlled trials. *Journal of Pediatrics* **149**: 367-372.

Takamine, F. and Imamura, T. 1995. Isolation and characterization of bile acid 7-dehydroxylating bacteria from human feces. *Microbiology and Immunology* **39**: 11-18.

Tankanow, R.M., Ross, M.B., Ertel, I.J., Dickinson, D.G., McCormick, L.S. and Garfinkel, J.F. 1990. A double-blind, placebo-controlled study of the efficacy of Lactinex in the prophylaxis of amoxicillin-induced diarrhea. *Drug Intelligence and Clinical Pharmacy* **24**: 382384.

Teasley, D.G., Gerding, D.N., Olson, M.M. *et al.* 1983. Prospective randomised trial of metronidazole versus vancomycin for *Clostridium difficile*-associated diarrhoea and colitis. *Lancet* **2**: 1043-1046.

Vanhoute, T., Huys, G., De Brandt, E. and Swings, J. 2004. Temporal stability analysis of the microbiota in human feces by denaturing gradient gel electrophoresis using universal and group-specific 16S rRNA gene primers. *FEMS Microbiology Ecology* **48**: 437-446.

Vaughan, E.E., Schut, F., Heilig, H.G., Zoetendal, E.G., de Vos, W.M. and Akkermans, A.D. 2000. A molecular view of the intestinal ecosystem. *Current Issues in Intestinal Microbiology* **1**: 1-12.

Viscidi, R., Willey, S. and Bartlett, J.G. 1981. Isolation rates and toxigenic potential of *Clostridium difficile* isolates from various patient populations. *Gastroenterology* **81**: 5-9.

von Eichel-Streiber, C., Laufenberg-Feldmann, R., Sartingen, S., Schulze, J. and Sauerborn, M. 1990. Cloning of *Clostridium difficile* toxin B gene and demonstration of high N-terminal homology between toxin A and B. *Medical Microbiology and Immunology* **179**: 271-279.

von Eichel-Streiber, C., Laufenberg-Feldmann, R., Sartingen, S., Schulze, J. and Sauerborn, M. 1992a. Comparative sequence analysis of the *Clostridium difficile* toxins A and B. *Molecular and General Genetics* **233**: 260-268.

Wenisch, C., Parschalk, B., Hasenhundl, M., Hirschl, A.M. and Graninger, W. 1996. Comparison of vancomycin, teicoplanin, metronidazole, and fusidic acid for the treatment of *Clostridium difficile*-associated diarrhea. *Clinical Infectious Diseases* **22**: 813-818.

Wistrom, J., Norrby, S.R., Myhre, E.B. *et al.* 2001. Frequency of antibiotic-associated diarrhoea in 2462 antibiotic-treated hospitalized patients: a prospective study. *Journal of Antimicrobial Chemotherapy* **47**: 43-50.

Wullt, M., Hagslatt, M.L. and Odenholt, I. 2003. *Lactobacillus plantarum* 299v for the treatment of recurrent *Clostridium difficile*-associated diarrhoea: a double-blind, placebo-controlled trial. *Scandinavian Journal of Infectious Diseases* **35**: 365-367.

Wullt, M. and Odenholt, I. 2004. A double-blind randomized controlled trial of fusidic acid and metronidazole for treatment of an initial episode of *Clostridium difficile*-associated diarrhoea. *Journal of Antimicrobial Chemotherapy* **54**: 211-216.

Wullt, M., Johansson Hagslatt, M.L., Odenholt, I. and Berggren, A. 2007. *Lactobacillus plantarum* 299v enhances the concentrations of fecal short-chain fatty acids in patients with recurrent *Clostridium difficile*-associated diarrhea. *Digestive Diseases and Sciences* **52**(9). 2082-2086.

Wunderlich, P.F., Braun, L., Fumagalli, I. *et al.* 1989. Double-blind report on the efficacy of lactic acid-producing *Enterococcus* SF68 in the prevention of antibiotic-associated diarrhoea and in the treatment of acute diarrhoea. *Journal of International Medical Research* **17**: 333-338.

Xu, J. and Gordon, J.I. 2003. Honor thy symbionts. *Proceedings of the National Academy of Sciences USA* **100**: 10452-10459.

Yearsley, K.A., Gilby, L.J., Ramadas, A.V., Kubiak, E.M., Fone, D. L. and Allison, M. C. 2006. Proton pump inhibitor therapy is a risk factor for *Clostridium difficile*-associated diarrhoea. *Alimentary Pharmacology and Therapeutics* **24**: 613-619.

Young, G.P., Ward, P.B., Bayley, N. *et al.* 1985. Antibiotic-associated colitis due to *Clostridium difficile*: double-blind comparison of vancomycin with bacitracin. *Gastroenterology* **89**: 1038-1045.

Young, R.J., Whitney, D.B., Hanner, T.L., Antonson, D.L., Lupo, J.V. and Vanderhoof, J.A. 1998. Prevention of antibiotic-associated diarrhea utilizing *Lactobacillus* GG. *Gastroenterology* **114**: A435.

Zoetendal, E.G., Akkermans, A.D.L. and de Vos, W.M. 1998. Temperature gradient gel electrophoresis analysis from human fecal samples reveals stable and host-specific communities of active bacteria. *Applied and Environmental Microbiology* **64**: 3854-3859.

Zoetendal, E.G., Akkermans, A.D.L., van-Vliet Akkermans, W.M., de Visser, J.A.G.M., and de Vos, W.M. 2001. The host genotype affects the bacterial community in the human gastrointestinal tract. *Microbial Ecology in Health and Disease* **13**: 129-134.

Responsible exploitation of scientific data on pro- and prebiotics for consumer health: a consensus document

Aat M. Ledeboer[1] and Liisa Lähteenmäki[2]

[1]*Lactic Acid Bacteria Industrial Platform (LABIP) and Unilever R&D, PO Box 114, 3130AC Vlaardingen, the Netherlands. Tel. 31-10-4605239. Fax 31-10-4605188. E-mail aat.ledeboer@unilever.com*

[2]*VTT Technical Research Centre of Finland, PO Box 1000, FI-02044 VTT, Finland*

Workshop participants: Barbara Gallani (The European Consumers' Organisation, BEUC, Belgium), Jarmila Hubená (Consumers Defence Association, Czech Republic), Liisa Lähteenmäki (VTT, Finland), Esben Laulund (Chr. Hansen, Denmark), Peter Leathwood (Nestle, Switzerland), Irene Lenoir-Wijnkoop (Danone Research Institute, France), Rick Mattes (Purdue University, USA), Rod Mitchell (EFCCA, UK), Mary Petri (Valio, Finland), Jarmo Saarikivi (Finnish Consumers Association), Seppo Salminen (University of Turku, Finland), Joachim Scholderer (MAPP, Denmark), Epp Songisepp (Biocompetence Centre of Healthy Dairy Products Ltd, Tartu, Estonia), Elaine Vaughan (Unilever R&D, the Netherlands). Science writer Paul Wymer, UK.

Abstract

As part of a European Union (EU) funded Specific Support Action called 'GutImpact', a consensus workshop was held in Rome in September 2005 where the consumer perspective on probiotics was discussed by representatives of industry, academia and consumer groups. The consensus document states that products need to fulfil any claims made and that these must be communicated by clear but simple health-related messages. This paper is the consensus document of a second workshop, held in September 2006 in Tallinn, Estonia. In this workshop, again attended by representatives from academia, industry and consumer groups, some of the themes of the earlier consensus document were developed and discussions concentrated on how the responsibilities related to promises and clarity of messages are divided among the different stakeholders (industry, authorities and consumers) in the chain. It was also discussed whether any special considerations for vulnerable groups will be needed, and hence what the implications might be for marketing and health claims.

Keywords: probiotics, prebiotics, consumer, health, legislation, claims

1. Introduction

Probiotics are defined by the World Health Organisation (WHO) as "live microorganisms which, when administered in adequate amounts, confer a health benefit on the host" (FAO/WHO 2002). In addition, products containing ingredients claimed to be beneficial for health, such as prebiotics—non-digestible food ingredients that beneficially affect the host by selectively stimulating the growth and/or activity of one or a limited number of bacteria in the colon and thus improve host health—form an attractive health-promoting proposition for the consumer (Salminen *et al.* 1998). Recent scientific evidence supports both concepts, although there are still gaps in our knowledge. To fulfil the promise of these foods as healthy options for consumers, industry has a clear obligation to use health claims responsibly and to substantiate their efficacy in cooperation with the authorities. This will make it possible for the consumer to make informed choices on the use of such foods.

As part of an EU-funded Specific Support Action called 'GutImpact' a consensus workshop was held in Rome in September 2005 where the consumer perspective on probiotics was discussed by representatives of industry, academia and consumer groups. The consensus document states that products need to fulfil any claims made and that these must be communicated by clear and simple health-related messages (Lähteenmäkki and Ledeboer 2006). This paper is the consensus document of a second workshop, held in September 2006 in Tallinn, Estonia. In this workshop, again attended by representatives from academia, industry and consumer groups, some of the themes of the earlier consensus document were developed, and discussions concentrated on how the responsibilities related to promises and clarity of messages are divided among the different stakeholders in the chain. At the workshop,

Food Science and Technology Bulletin: Functional Foods 4 (5) 43–49
DOI: 10.1616/1476-2137.14991. Published 22 October 2007
ISSN 1476-2137

participants were divided into subgroups to discuss the following topics:

- Special considerations for vulnerable groups and the role of the consumers' own understanding with regard to implications for marketing and health claims.
- The responsibility and roles of stakeholders in assessing the efficiency and suitability of gut health products for individual consumers
- The acceptability and trustworthiness of different marketing practices, including advertising and health claims

The outcomes of the discussions were presented to the whole group, and comments were further discussed in the subgroups. After the workshop, this consensus document was developed by the interactive exchange of comments and text, until all participants had agreed.

2. Probiotics and prebiotics with proven health effects: are there vulnerable groups and how should the health message be formulated?

Fermented products containing living microorganisms have traditionally been used to restore gut health. Probiotics have been defined as live microorganisms which, when administered in adequate amounts, confer a health benefit on the host (FAO/WHO 2002). Selected strains, mainly belonging to the genera *Lactobacillus* and *Bifidobacterium*, are increasingly being used as probiotics. The concept of prebiotics concerns the use of carbohydrate components that are not absorbed in the upper gastrointestinal tract, but are selectively fermented in the colon, thus stimulating positive changes in microbiota and resulting in health benefits for the host (Salminen *et al.* 1998).

Recent work has provided a mechanism for the proposed health effects of probiotics (Guarner and Malagelada 2003), while human intervention studies have demonstrated that a significant reduction in the risk of gut-related diseases can be achieved through the use of probiotics in specific human populations (Ouwehand *et al.* 2003). Prebiotics have been used in human studies and in a recent review both the benefits and the need for further research were emphasised (Macfarlane *et al.* 2006). The probiotic concept has now gained widespread acceptance and the prebiotic concept is becoming so, especially through demonstrated health effect in studies in infants (Bakker-Zierikzee *et al.* 2005; Moro *et al.* 2006).

2.1 Importance of the intestinal microbiota

Understanding the crosstalk that occurs between intestinal microbiota and the host is necessary in elucidating how the former impacts on health (Backhed *et al.* 2005). There is increasing evidence that specific aberrations in the intestinal microbiota may make the host more vulnerable to intestinal inflammatory diseases and to abnormalities in other organs (Dumas *et al.* 2006). It is likely that some abnormalities even predispose us to specific diseases. While there is increasing knowledge of the composition of the intestinal microbiota and factors that govern it, there is still much to be understood about its diversity and, in particular, its functionality and how it interacts with the host. Assessment of the effects of probiotic consumption on the microbiota may be important for selection and evaluation of new probiotics with specific targets.

2.2 Probiotic effects

Due to their unique interaction with the host, the data concerning the health benefits of the most common probiotics needs to be assessed on a strain by strain basis before any health-related product claims can be approved. In general, probiotics do not permanently colonise the human intestinal tract, but some strains are able to colonise transiently and to modulate the indigenous microbiota. Some may also modulate local and systemic immune responses (Isolauri *et al.* 2002). Other mechanisms of probiotic action have been proposed, such as inhibition of pathogens by competition for nutrients and attachment sites or by the production of antimicrobial substances, reduction of cholesterol levels through deconjugation of bile salts, or the binding of toxins and carcinogens, precluding their absorption (Mercenier *et al.* 2003). Genomic data for some members of the human intestinal microbiota indicates how these bacteria are adapted to the gut (Xu *et al.* 2003). Genomic research has also provided information about the adhesive properties of probiotic microorganisms, which suggests mechanisms both for populating the gut and for communicating developmental signals to specific areas of the gut mucosa (Kleerebezem and Hugenholtz 2003).

Knowledge of genome sequences provides an insight into the potential properties of these microorganisms but not into the way they might behave *in vivo*. It is possible that further genomic research, such as DNA microarrays, might unravel the *in vivo* functions of probiotics (Vaughan *et al.* 2005) and allow monitoring of their effects on host gene expression (Di Caro *et al.* 2005). Application of efficacy assessment protocols normally used in the pharmaceutical industry has provided a standard for probiotic studies. Such investigations suggest that certain probiotics have proven benefits. Several meta-analyses identify the treatment and prevention of acute diarrhoea and antibiotic-associated side-effects as being the best documented health effects (D'Souza *et al.* 2002; Van Niel *et al.* 2002; Sazawal *et al.* 2006). Other reported probiotic health-related effects, such as the reduction of certain allergies or the symptoms of irritable bowel syndrome are only partially established and require further investigation in the form of larger, double-blind, placebo-controlled studies.

2.3 Prebiotic effects

The interest in prebiotics focussed initially on their bifidogenic effect (i.e. the increased number of bifidobacteria in the colon) and fibre-like characteristics, which could alleviate digestive symptoms. However, it remains to be confirmed that an increase in bifidobacteria by prebiotics provides a health benefit for the host. On the other hand, bifidobacteria are the dominant commensal microbiota of most breast-fed babies, and the negative effects seen upon delayed colonisation of bifidobacteria in pre-term babies have been documented (Westerbeek *et al.* 2006). The underlying assumption is that at least part of the protective effect of breast-feeding is linked to the presence of bifidobacteria. Lately, the interest in prebiotics has also focussed on lipid and mineral metabolism. The gut is the largest immune organ in the body and the intestinal microbiota is known to have a profound impact in terms of modulating immunity and shaping the immune system from a young age. Thus, prebiotics that have an influence on the diversity and metabolic activity of the microbiota may consequently impact on the immune system.

A prebiotic index approach has been devised, which uses *in vitro* gut models and involves the enumeration of four major microbiota groups (Palframan *et al.* 2003). The index was designed to aid in the quantification of a prebiotic effect and hence should reflect what is occurring in the *in vivo* situation; but so far this has not been confirmed for all prebiotics. Multidisciplinary research is required to unravel all of the mechanisms of action of prebiotics, some of which are likely to involve synergism with other prebiotics or with probiotics. A greater understanding of the gut microbiota and recent advances in 'omics' technologies are expected to help unravel mechanisms of action of prebiotics and to enable the design of new prebiotics.

2.4 Selection of probiotics and prebiotics

A reliable probiotic product requires correct identification of the bacterial species used and an indication of these on the product label. This is important as a number of recent reports have shown that the identity of recovered microorganisms does not always correspond to the information stated on the product label (Temmermen *et al.* 2003). To produce therapeutic benefits, a sufficient number of viable microorganisms or a sufficient dose of a particular prebiotic component must be present in the product. The basis for assessing probiotic efficacy in humans requires an understanding of probiotic strains, each of which is unique. In the same way, each prebiotic component is unique and influences a specific part of the colonic microbiota. Prebiotics such as inulin and fructooligosaccharides occur naturally in many foodstuffs. Thus, strain properties and prebiotic characteristics should be well-defined and studies using closely related probiotic strains or prebiotics cannot be extrapolated to support others.

2.5 Safety of probiotics and prebiotics

An assessment of safety is essential in the selection and evaluation of probiotics. Few probiotic strains have been specifically tested, but the long history of safe consumption of some probiotics could be considered proof of their safety. Although some lactobacilli and bifidobacteria have been associated with rare cases of bacteraemia, usually in patients with severe underlying diseases, the safety of members of these genera is generally recognised due to their history of safe use. The low incidence of infections attributable to these microorganisms, together with recent studies in Finland or Sweden, showing that there are no increases in the incidence of bacteraemia due to lactobacilli despite the increased consumption of probiotic lactobacilli (Salminen *et al.* 2002; Sullivan and Nord 2006), supports this hypothesis. Furthermore, even clinical trials using patients suffering from serious intestinal conditions, including inflammatory bowel diseases such as ulcerative colitis and Crohn's disease, indicate their safety (Sheil *et al.* 2007).

All new potential probiotic strains should undergo safety testing, the level of which will depend on whether the species already has GRAS ('generally regarded as safe') status—which applies to most members of the lactic acid bacteria group and human bifidobacteria—and whether they are species that are commonly found in healthy gut microbiota. Currently it is not proven that a microbe is more likely to be beneficial and effective as a probiotic if isolated in humans; however, probiotics from elsewhere exist, and new ones are not ruled out. Other bacteria applied as probiotics, such as enterococci, *Clostridium butyricum* some members of the genus *Bacillus* or the yeast *Saccharomyces boulardii*, require further case by case assessment.

Ingestion of large amounts of prebiotics can in certain cases cause flatulence, abdominal cramps and even diarrhoea. This is not really a safety issue, but more an inconvenience and most people adapt to the prebiotics. One particular issue was reported in rats where specific prebiotics appeared to cause a change that enhanced colonisation and translocation of an intestinal pathogen *Salmonella* in low calcium conditions (Ten Bruggencate *et al.* 2004). Such observations have not been reported in humans or in other animal models where similar prebiotics have been shown to have beneficial effects (Guarner 2006; Ten Bruggencate *et al.* 2006).

3. Responsibility and roles of stakeholders in assessing efficiency and suitability of gut health products for individual consumers

3.1 Authorities

Overall principles and definitions should be regulated at the EU or international level (FAO/WHO; Codex), but local interpretation of regulations is necessary to ensure that an allowance is made for the local culture. However, in order to avoid confusion, independent umbrella organisations should set definitions and, if relevant, recommendations for minimum and/or maximum amounts of the ingested ingredients or components, and these should be accepted as the general standard, both on international and local levels. In Europe, the role of the EU is to provide rules governing what are appropriate and acceptable claims. The on-going project to establish a definitive list of acceptable generic claims as part of EU legislation is welcomed. Specific claims related to a reduction in the risk of disease will be evaluated on a case by case basis. For consumers, this ensures that these health claims are based on solid evidence. If evidence is considered weak or is still unsure, no claim should be allowed. Also, different levels of proof (level A, B or C) are not recommended in communicating health benefits to the consumer.

The cultural suitability of claims needs to be decided at a national level. This applies both to the wording and the evaluation of the dietary role of products, and their relevance to local public health concerns. As similar expressions have different connotations in different languages, the wording, use of conditional terms and level of specificity has to be validated and endorsed at a national level although, whenever possible, a common set of terms should be used to prevent confusion for the consumer. Furthermore, national authorities should be able to make supplementary and binding guidelines on how regulations should be interpreted. Finally, local exemptions should be possible when authorities perceive an urgent need for this. Local authorities should carry out sufficient market surveillance on pro- and prebiotic market sectors to assess how products fulfil their claims. Such monitoring will ensure responsible behaviour on the behalf of providers and thus justify and promote consumer confidence.

The responsibility for informing consumers about the benefits of single products lies with the producers, but there is no clear view on the most appropriate body to provide information about the general role of pro- and prebiotics products in people's health and well-being. Possibilities for this include health professionals involved in nutrition education, consumer organisations, and the food industry or food authorities. Authorities should be proactive in engaging in dialogue with stakeholders to establish the responsible role of industry as well as consumer trust in products.

3.2 Industry

The responsibility of industry for its products includes an assessment of the efficacy of the bioactive ingredient, suitability of the food matrix and a guarantee of safety, as for any food product. Industry is also responsible for providing sufficient information about the benefits of the product in an understandable form, avoiding highly technical medical terms or names of chemical compounds that have little consumer relevance. The science-based health claim must be made in a form that is distinct from other, more commercially driven information.

Consumers should be informed about the serving required to gain the putative health effect, and about appropriate use, such as storage conditions needed to retain optimal activity or, in case of foods that might be cooked, information about any potential loss of benefit.

If the product is beneficial only to a particular target group, this should be clearly stated. There should be no health-related risk to any consumer group if they use the product, even if not being targeted, but the target and non-target groups should be able to recognise themselves so that consumers are not misled. Industry is responsible for monitoring products after they have been launched, and post-market monitoring should include an evaluation of product quality and efficacy throughout its shelf life and also the way it is used by consumers, as well as any possible inconveniences or undesirable effects. Industry should agree further on how to monitor whether the consumers do receive the benefits promised by the product. Product-related information should be modified as necessary in the light of this evidence. At all times, industry should be ready to participate in open dialogue and provide information about the benefits and any side-effects their products may have. Consumers are more and more actively involved in product development, including translation of the scientifically proven benefit into meaningful information for other consumers. Further possibilities for consumer input into innovation and post-market monitoring should be explored.

3.3 Individual consumers

Consumers have responsibility for their own food choices. The 'average' consumer is regarded as a person who has a 'normal' capability for making these decisions about his or her own choices. However, when providing health-related information, product by product considerations should be made to assess whether there are any possible risks to specific target groups, such as children, the elderly or those diagnosed with chronic diseases. Product-specific information provided by industry and authorities should be in-line with the regulatory framework and rules for the pro- and prebiotic food market. It is the consumers' responsibility to compare various products available, to be able to make an informed choice. Consumers should also be ready to actively participate in dialogue and provide feedback in the experiences and issues that are important from the consumers' point of view. Improved methods and systems to facilitate this process should be developed.

4. Acceptability and trustworthiness of different marketing practices, including advertising and claims

4.1 BEUC survey on consumers' perceptions of foodstuffs labelling

In September 2005 the European Consumers' Organisation BEUC, published a survey of consumers' perceptions of foodstuffs labelling in five European countries: Germany, Denmark, Spain, Hungary and Poland (BEUC 2005). Three thousand individuals were interviewed and of these approximately 75% were women; 70% of respondents were aged between 20 and 59 years, with 28% over 60. Just over 50% of the respondents were classed as being of normal weight (BMI: 18–25), 35% were overweight and 12% were obese. The level of interest in nutrition was generally quite high, with 75% respondents declaring themselves to be very or sometimes interested in nutrition.

The survey examined consumers' use of labelling such as: 'low in fat', 'low in sugar', 'high in vitamins/minerals' etc.; 59% of interviewees declared that such claims always or often caught their attention and that they read them. Only 14% said that they did not read claims, compared to over 25% of respondents who never read the nutritional table. More than 50% of interviewees agreed that a claim could lead them to buy a product. The effectiveness of nutritional claims in attracting consumers' attention and influencing their behaviour was very clear: 80% of interviewees indicated that claims were easy to locate, and 70% believed they were easy to understand. In addition, over 50% of interviewees said that they trusted claims.

Consumer trust in nutrition claims is generally associated with their trust in the brand (65%), and in the retailer (18%). This, therefore, puts a great onus on the brand owner and retailer to ensure that claims on products are true and do not mislead the consumer. Interestingly, those who say that they do not trust claims also admit to being influenced by them. It is important to note that many interviewees believed that they understood the meaning of a nutrition claim and admitted being influenced by it, even if they had provided incorrect answers to questions on nutrition.

4.2 Scientific substantiation of health claims

In-line with the influence of health and nutrition claims on consumer trust, only health claims with a high level of scientific substantiation should be allowed, and the existence of an independent authorisation procedure (by way of the European Food Safety Authority) for assessing the scientific basis for claims will help to increase consumer trust. Claims should only be accepted in accordance with the new EU regulation on nutrition and health claims made on foods (European Union 2007). The relevant paragraphs are as follows:

"Scientific substantiation should be the main aspect to be taken into account for the use of nutrition and health claims and the food business operators using claims should justify them."

As the legislation points out:

"At international level the Codex Alimentarius has adopted General Guidelines on Claims in 1991 and Guidelines for the Use of Nutrition Claims in 1997. An amendment to the latter has been adopted by the Codex Alimentarius Commission in 2004. That amendment concerns the inclusion of health claims in the 1997 Guidelines. Due consideration is given to the definitions and conditions set in the Codex Guidelines."

For scientific substantiation of health claims, Article 6 of the proposed legislation is straightforward and states:

1. "Nutrition and health claims shall be based on and substantiated by generally accepted scientific evidence.
2. A food business operator making a nutrition or health claim shall justify the use of the claim.
3. The competent authorities of the Member States may request a food business operator or a person placing a product on the market to produce the scientific work and the data establishing compliance with this regulation."

Paragraph 23 of the introduction provides an outline as to how this should be interpreted:

"Health claims should only be authorised for use in the community after a scientific assessment of the highest possible standard. In order to ensure harmonised scientific assessment of these claims, the European Food Safety Authority should carry out such assessments. Upon request the applicant should be able to have access to his file to check the state of the procedure."

In this context it is worth mentioning that the European Commission has funded valuable projects such as the Concerted Action PASSCLAIM, aimed at setting principles for assessing the scientific support of health claims, and that this considerable work should be taken into account when assessing claims (Aggett *et al.* 2005). As stated above, communication of claims to the final consumer is very important. Therefore, to ensure that health claims are truthful, clear and reliable, the authority and the subsequent authorisation procedure, should take into account the wording of the claims assessed.

4.3 Consumer understanding of health claims

Consumers cannot be expected to have the knowledge necessary to understand the science behind health claims relating to probiotics (or other products). Instead, as stated in Article 5 of the proposed legislation, the use of nutrition and health claims should only be permitted if "the average consumer can be expected to understand the beneficial effects as expressed in the claim." In addition, pictures and symbols could be useful aids to communication as long as they are not misleading. When the new EU health claim legislation comes into force, consumer education may be needed to explain what these regulations require (i.e. that documentation for all accepted claims is officially evaluated).

5. Conclusions

Human studies have demonstrated that specific probiotic strains can have health benefits in specific human populations. Similarly, the results of human studies are emerging for prebiotics, although meta-analysis indicates a clear need for further work. Such evidence has led to wide acceptance of the probiotic and prebiotic concepts. Increased knowledge of intestinal microbiota, nutrition, immunity and mechanisms of probiotic action can now be combined with genomic data enabling isolation and characterisation of target- or disorder-specific probiotics. Similar developments may take place with specific prebiotic combinations, which will impact on safety assessment. However, no relevant vulnerable risk groups have been identified to date, and safety assessment studies support the overall safety of probiotic and prebiotic products on the market. Reported adverse effects with probiotics are few, and over dosing does not appear to be an issue. However, there are consumer groups such as inflammatory bowel disease patients that are more sensitive and vulnerable to gastrointestinal disturbances when large amounts of prebiotics are consumed. It is therefore important to indicate quantities of prebiotic ingredients contained within foods with any health messages related to them, to ensure appropriate doses and to minimise side-effects.

The consumer must be able to assume that an adequate dose of a probiotic is present in the product to provide the beneficial effect, which should be clearly indicated. Supplementary information should be available on internet sites or other corporate media that can further explain the beneficial effects and product messages. The dose, frequency of use and limits for consumption without side-effects should be taken into account when designing the marketing messages and product health claims for probiotics and prebiotics. Since food products containing pro- and prebiotics are intended for everyday consumption in typical quantities then no specific warnings are needed as, by definition of being a food, they should be safe for everyone. However, it is important to define the term 'average consumer' clearly and systematically. For example, does it include all people who buy food products, or only those interested in the product that contains a health-related claim? The implication that the average consumer is anyone who buys food products, suggests that health claims have to convey relevant information, whether the individual making the purchasing decision can benefit from the product or not. In this case, the message should state what the benefit is in clear and simple terms. As some products are beneficial only to certain groups, e.g. those with elevated cholesterol levels or high blood pressure, or the elderly etc, the need to define a specified target population has to be considered on a case by case basis.

In summary, in addition to the health claim, the following information should be provided on probiotic labels: specification of the bacteria included; the serving size necessary to be effective; information regarding the proper storage conditions and use; and relevant corporate contact details. In addition, the following information could be included on packaging or made readily available through other media, such as an easily accessible website, or on request from the corporate contact: a description of the physiological effect of the probiotic; the number of bacteria necessary to be effective; and the number of each strain remaining at the end of a standard shelf life.

The EU legislation deals with labelling presentation as follows:

"General labelling provisions are contained in Directive 2000/13/EC of the European Parliament and of the Council of 20 March 2000 on the approximation of the laws of the Member States relating to the labelling, presentation and advertising of foodstuffs, as amended by Commission Directive 2001/101/EC. Directive 2000/13/EC generally prohibits the use of information that would mislead the purchaser or attribute medicinal properties to food. This Regulation should complement the general principles laid down in Directive 2000/13/EC and lay down specific provisions concerning the use of nutrition and health claims concerning foods to be delivered as such to the consumer."

Agencies involved in assessing and proving the efficiency and suitability of pro- and prebiotics are defined above as authorities, industry and individual consumers making their food purchases. However, it is recognised that healthcare professionals (medical and nutritional) are important providers of health- and diet-related advice. Their need for information about pre- and probiotics and for understanding connections between these types of new food products and health should gain more attention. These groups are also partners in providing the scientific substantiation for the health claims.

6. Acknowledgements

The workshop was supported financially by the EU funded Specific Support Action GutImpact FP6-2003-517821.

7. References

Aggett, P. J., Antoine, J. M., Asp, N. G. *et al.* 2005. PASSCLAIM: process for the assessment of scientific support for claims on food; consensus on criteria. *European Journal of Nutrition* **44** (Suppl. 1): 1-30.

Backhed, F., Ley, R. E., Sonnenburg, J. L. *et al.* 2005. Host-bacterial mutualism in the human intestine. *Science* **307**: 1915-1920.

Bakker-Zierikzee, A. M., Alles, M. S., Knol, J. *et al.* 2005. Effects of infant formula containing a mixture of galacto- and fructo-oligosaccharides or viable *Bifidobacterium animalis* on the intestinal microflora during the first 4 months of life. *British Journal of Nutrition* **94**: 783-790.

BEUC. 2005. *Report on European consumers' perception of foodstuffs labelling. Results of consumer research conducted on behalf of BEUC from February to April 2005.* BEUC/X/032/2005.

Di Caro, S., Tao, H., Grillo, A. *et al.* 2005. *Bacillus clausii* effect on gene expression pattern in small bowel mucosa using DNA microarray analysis. *European Journal of Gastroenterology and Hepatology* **17**: 951-960.

D'Souza, A., Chakravarthi, R., Cooke, J. and Bulpitt, C. J. 2002. Probiotics in prevention of antibiotic associated diarrhoea: meta-analysis. *British Medical Journal* **324**: 1361-1366.

Dumas, M. E., Barton, R. H., Toye, A. *et al.* 2006. Metabolic profiling reveals a contribution of gut microbiota to fatty liver phenotype in insulin-resistant mice. *Proceedings of the National Academy of Sciences USA.* **103**: 12511-12516.

European Union. 2007. *Regulation (EC) No. 1924/2006 of the European Parliament and of the Council of 20 December 2006 on nutrition and health claims made on foods.* (Corrigendum: 18.1.2007).

FAO/WHO. 2002. *Joint FAO/WHO working group report on drafting guidelines for the evaluation of probiotics in food.* 30 April–1 May, 2002, London, UK and Ontario, Canada.

Guarner, F. 2006. Letters to the Editor: prebiotics and mucosal barrier function. *Journal of Nutrition* **136**: 2269.

Guarner, F. and Malagelada, J. R. 2003. Role of bacteria in experimental colitis. *Best Practice and Research in Clinical Gastroenterology* **17**: 793-804.

Isolauri, E., Kirjavainen, P. V. and Salminen, S. 2002. Probiotics: a role in the treatment of intestinal infection and inflammation? *Gut* **50** (Suppl. 3): 54-59.

Kleerebezem, M. and Hugenholtz, J. 2003. Metabolic pathway engineering in lactic acid bacteria. *Current Opinion in Biotechnology* **14**: 232-237.

Lähteenmäki, L. and Ledeboer, A. M. 2006. Probiotics – the consumer perspective. *Food Science and Technology Bulletin* **3**: 47-50.

Macfarlane, S., Macfarlane, G. T. and Cummings, J. H. 2006. Prebiotics in the gastrointestinal tract. *Alimentary Pharmacology and Therapeutics* **24**: 701-714.

Mercenier, A., Pavan, S. and Pot, B. 2003. Probiotics as biotherapeutic agents: Present knowledge and future prospects. *Current Pharmaceutical Design* **9**: 175-191.

Moro, G., Arslanoglu, S., Stahl, B. *et al.* 2006. A mixture of prebiotic oligosaccharides reduces the incidence of atopic dermatitis during the first six months of age. *Archives of Disease in Childhood* **91**: 814-819.

Ouwehand, A. C., Salminen, S., Roberts, P. J. *et al.* 2003. Disease-dependent adhesion of lactic acid bacteria to the human intestinal mucosa. *Clinical and Diagnostic Laboratory Immunology* **10**: 643-646.

Palframan, R., Gibson, G. R. and Rastall, R. A. 2003. Development of a quantitative tool for the comparison of the prebiotic effect of dietary oligosaccharides. *Letters in Applied Microbiology* **37**: 281-284.

Salminen, S., Bouley, C., Boutron-Ruault, M. C. *et al.* 1998. Functional food science and gastrointestinal physiology and function. *British Journal of Nutrition* **80**: S147-S171.

Salminen, M. K., Tynkkynen, S., Rautelin, H. *et al.* 2002. *Lactobacillus* bacteremia during a rapid increase in probiotic use of *Lactobacillus rhamnosus* GG in Finland. *Clinical Infectious Diseases* **35**: 1155-1160.

Sazawal, S., Hiremath, G., Dhingra, U. *et al.* 2006. Efficacy of probiotics in prevention of acute diarrhea: a meta-analysis of masked, randomized, placebo-controlled trials. *Lancet Infectious Diseases* **6**: 374-382.

Sheil, B., Shanahan, F. and O'Mahony, L. 2007. Probiotic effects on inflammatory bowel disease. *Journal of Nutrition* **137**: 819S-824S.

Sullivan, A. and Nord, C. E. 2006. Probiotic lactobacilli and bacteraemia in Stockholm. *Scandinavian Journal of Infectious Diseases* **38**: 327-331.

Temmerman, R., Pot, B., Huys, G. and Swings, J. 2003. Identification and antibiotic susceptibility of bacterial isolates from probiotic products. *International Journal of Food Microbiology* **81**: 1-10.

Ten Bruggencate, S. J., Bovee-Oudenhoven, I. M., Lettink-Wissink, M. L., Katan, M. B. and van der Meer, R. 2004. Dietary fructo-oligosaccharides and inulin decrease resistance of rats to *Salmonella*: protective role of calcium. *Gut* **53**: 530-535.

Ten Bruggencate, S. J., Bovee-Oudenhoven, I. M., Lettink-Wissink, M. L., Katan, M. B. and van der Meer, R. 2006. Dietary fructooligosaccharides affect intestinal barrier function in healthy men. *Journal of Nutrition* **136**: 70-74.

Van Niel, C. W., Freudtner, C., Garrison, M. M. and Christakis, D. A. 2002. *Lactobacillus* therapy for acute infectious diarrhea in children: a meta-analysis. *Pediatrics* **109**: 678-684.

Vaughan, E. E., Heilig, H. G., Ben-Amor, K. and de Vos, W. M. 2005. Diversity, vitality and activities of intestinal lactic acid bacteria and bifidobacteria assessed by molecular approaches. *FEMS Microbiology Reviews* **29**: 477-490.

Westerbeek, E. A., van den Berg, A., Lafeber, H. N. *et al.* 2006. The intestinal bacterial colonisation in preterm infants: a review of the literature. *Clinical Nutrition* **25**: 361-368.

Xu, J., Bjursell, M. K., Himrod, J. *et al.* 2003. A genomic view of the human – *Bacteroides thetaiotaomicron* symbiosis. *Science* **299**: 2074-2076.

Risk management of vitamins and minerals: a risk categorisation model for the setting of maximum levels in food supplements and fortified foods

David P. Richardson

DPR Nutrition Ltd, 34 Grimwade Avenue, Croydon, Surrey CR0 5DG, UK.

Abstract

The provisions of the European regulation 1925/2006, which harmonise the conditions for the voluntary addition of vitamins and minerals to foodstuffs (food fortification), and of the Directive 2002/46/EC, which relate to food supplements, both set out criteria that must be taken into account when establishing the maximum amounts of these essential nutrients that are present in products. The current paper reviews the application of the principles of risk management in order to underpin these regulatory developments. It takes into account international risk assessments of tolerable upper intake levels (ULs) of vitamins and minerals, and current market practices in Europe. A methodology is proposed that characterises risk using a 'population safety index' and thus provides a process by which quantitative and qualitative data can be used to allocate the nutrients into three categories of risk. The proposed risk management model is also used as an aid to the setting of maximum levels of vitamins and minerals in fortified foods and food supplements.

Keywords: EU regulation, maximum levels, vitamins, minerals, food supplements, food fortification

1. Introduction

There are three major and complementary ways of delivering the essential micronutrients required for human health and well-being:

- By promoting the consumption of nutrient-dense foods
- By increasing the availability and intake of foods with added nutrients
- By the appropriate use of food supplements.

Whereas all of these approaches have been demonstrated to provide safe ways of ensuring the nutritional status of populations and individuals at certain times in their lives (World Bank 1994; Food and Agriculture Organisation and International Life Sciences Institute 1997; Micronutrient Initiative 2007), it is necessary for the scientific community, regulatory authorities and industry to work together to ensure that levels of micronutrients in the total diet are safe, and that the sum of intakes from all the sources does not lead to excessive intakes and any adverse effects.

Food Science and Technology Bulletin: Functional Foods 4 (6) 51–66
DOI: 10.1616/1476-2137.14996. Published 2 November 2007
ISSN 1476-2137

The recent European Commission (EC) orientation paper on setting maximum and minimum amounts for vitamins and minerals in foodstuffs (EC 2007) highlights the criteria to be taken into account for the establishment of the maximum amounts present in food supplements and fortified foods. The EC orientation paper summarises first the provisions, laid down in Regulation 1925/2006 (European Parliament and of the Council 2006), which harmonise the conditions for the voluntary addition of vitamins and minerals and of certain other substances to foods (often referred to informally as food fortification). Second, it summarises the provisions of Directive 2002/46/EC (European Parliament and Council 2002) on the approximation of the laws of member states relating to food supplements that harmonise specific rules on vitamins and minerals in these products. Both legal acts aim to provide a high level of consumer protection and to ensure the effective functioning of the internal market.

The key criteria for setting maximum amounts in foods and food supplements are:

- The upper safe levels of vitamins and minerals established by scientific risk assessment based on generally accepted scientific data, taking into account, as appropriate, the varying degrees of sensitivity of different consumer groups

- The intake of vitamins and minerals from all dietary sources
- Reference intakes of vitamins and minerals for the population.

When maximum amounts are set for vitamins and minerals in normal foods, whose reference intakes for the population are close to the upper safe levels, the following should also be taken into account:

- The contribution of individual products to the overall diet of the population in general, or of subgroups of the population
- The established nutrient profile of the product, as provided for by Regulation (EC) 1924/2006 (European Parliament and of the Council 2007).

The present paper reviews the application of the principles of risk management in order to underpin European and international regulatory developments relating to the setting of maximum levels of vitamins and minerals in food supplements and in foods with added nutrients. The review also takes into account current market practices in Europe and provides a process for the categorisation of nutrients according to the risk of exceeding the UL, as determined by either the Scientific Committee on Food and the Scientific Panel on Dietetic Products, Nutrition and Allergies (2006),[1] or by other internationally recognised scientific risk assessments.

2. Control of risk in foods and food components

Risk analysis comprises three distinct but closely linked components (Codex Alimentarius Commission 2003; World Health Organisation 2002, 2005): risk assessment, risk management and risk communication. Each component is integral to the overall risk analysis, and the primary objective is not only to protect the health of consumer, but also to take into account other legitimate factors including the facilitation of consumer choice and avoidance of barriers to international trade.

As indicated in the introduction, one of the criteria to be taken into account for the establishment of maximum amounts of vitamins and minerals in fortified foods and food supplements is the safe levels of vitamins and minerals established by scientific risk assessment. The EC requested the Scientific Committee on Food (SCF) and later the European Food Safety Authority (EFSA) to provide scientific opinions on ULs for 29 nutrients listed in Annex I of the Food Supplements Directive. This resulted in specific numerical UL being established for 16 nutrients. Some of the remaining nutrients showed extremely low or non-existent adverse effects, even at very high levels of intake, and for some, the lack of sufficient scientific data did not permit the derivation of a numerical UL. Where UL have not been established, SCF/EFSA have provided qualitative risk characterisations for the specific nutrients. In contrast, other expert scientific risk assessment committees, such as the UK Expert Group on Vitamins and Minerals (EVM 2003)[2] and the US Institute of Medicine (IOM 1999, 2000a, 2000b, 2002) have set numerical UL values and these have been taken into account on a case-by-case basis in the current risk management model. Table 1 shows the levels established by these three risk assessment committees (SCF/EFSA, EVM and IOM).

3. Qualitative risk characterisation of nutrients without a UL and with no evidence of risk to human health

Qualitative assessment of SCF/EFSA opinions and other authoritative reports, including those of the EVM and IOM, show no adverse effects in healthy individuals associated with high intakes of vitamin B_1 (thiamin), vitamin B_2 (riboflavin), vitamin B_{12} (cobalamin), biotin, panthothenic acid, chromium (trivalent form) and vitamin K. A summary of the qualitative risk characterisation of this group of nutrients (group A) is shown in Table 2, which demonstrates that these nutrients have extremely low or non-existent toxicity even at high levels of intake. In this context and in the absence of any adverse effects, there is no scientific basis for taking risk management measures. Although a lack of evidence of adverse effects should not be taken as evidence of absence, the establishment of maximum amounts for these nutrients in group A would seem to be disproportionate. The recent EC orientation paper (EC 2007) also concludes that, because of the lack of evidence of adverse effects a proportionate approach, in-line with the principles of better regulation, would be to not establish maximum amounts for these nutrients.

The EC orientation paper questions the positioning of two nutrients in group A, namely chromium and vitamin K. In the case of chromium, the biological effects depend strongly on its specific chemical form, and the nutritive

[1] UL defined as the maximum level of total chronic intake of a nutrient (from all sources) judged to be unlikely to pose a risk of adverse health effects to humans. 'Tolerable intake' in this context connotes what is physiologically tolerable and is a scientific judgement as determined by assessment of risk, i.e. the probability of an adverse effect occurring at some specified level of exposure. ULs may be derived for various lifestage groups in the population. The UL is an estimate of the highest level of intake that carries no appreciable risk of adverse health effects. To establish whether an exposed population is at risk requires a risk assessment to determine what is the fraction, if any, of the population whose intake exceeds the UL and the magnitude and duration of the excessive intake.

[2] The EVM risk assessments are based on the UL method, however the scientific committee assigned the term Safe Upper Level (SUL).

Table 1. A comparison of the upper safe levels for total daily intake from the Scientific Committee on Food (SCF) and the European Food Safety Authority (EFSA) and the US Institute of Medicine (IOM), and the daily levels for supplementation proposed by the UK Food Standards Agency Expert Group on Vitamins and Minerals (EVM)

Nutrient	Unit	SCF/EFSA total intake (UL)	IOM total intake (UL)	EVM for long-term supplementation (SUL)[a]
Vitamin A	μg	3000	3000	1500 (G, T)
β-carotene	mg	Not set	Not set	7 (not for smokers)
Vitamin D	μg	50	50	25 (G)
Vitamin E[b]	mg	300	1000	540 (800 IU)
Vitamin K	μg	Not set	Not set	1000 (G)
Thiamin (B_1)	mg	Not set	Not set	100 (G)
Riboflavin (B_2)	mg	Not set	Not set	40 (G) (43T)
Nicotinamide	mg	900	35[c]	500 (G) (560T)
Nicotinic acid	mg	10	–	17
Pantothenic acid	mg	Not set	Not set	200 (G) (210T)
Pyridoxine (B_6)	mg	25	100	200 (short term)[d] 10 (long term)
Folic acid	μg	1000 (+ dietary folate)	1000 supp. (+ 200 diet)	1000 (G) (1500T)
Vitamin B_{12}	μg	Not set	Not set	2000 (G)
Biotin	μg	Not set	Not set	900 (G) S (970T)
Vitamin C	mg	Not set	2000	1000 (G)
Calcium	mg	2500	2500	1500 (G)
Magnesium	mg	250 as supplement	350 as supplement+ diet	400 (G)
Iron	mg	Not set	45	17 (G)
Copper	mg	5	10	1 (10T)
Iodine	μg	600	1100	500 (G) (940T)
Zinc	mg	25	40	25 (42T)
Manganese	mg	Not set	11	4 (G) (9–12T) 0.5 (G) for older people
Potassium	mg	Not set	Not set	3700 (G)
Selenium	μg	300	400	350 (450T)
Chromium (trivalent)[e]	mg	Not set	Not set	10 (G, T)
Molybdenum	μg	600	2000	Not set
Fluoride	mg	Not set	10	Outside terms of reference
Phosphorus	mg	Not set	4000	250 (G) (2400T)

G, guidance level; T, total intake; IU, International Unit.
[a]All amounts relate to 60 kg bodyweight adult and figures in parentheses are total (T) amounts from all dietary sources.
[b]D-α-tocopherol equivalents/day.
[c]This UL is applied to the total of all forms of niacin resultant on the IOM's decision to establish a lowest-observed-adverse-effect level (LOAEL) based on skin flushing by nicotinic acid. In the EU niacin supplements and niacin fortification are generally in the form of nicotinamide.
[d]Implied in text of report.
[e]Picolinates are excluded.

effects are exclusively related to the trivalent form—chromium III. All of the major toxic effects are associated with chromium VI. No credible data or reports have demonstrated any adverse effects of chromium III in humans, and animal data also suggests that orally administered chromium is extremely innocuous. Overall, the risk assessors concluded that there is insufficient data to derive a UL for chromium. However, in their risk characterisation SCF/EFSA concluded that, in the limited number of human studies conducted, there was no evidence of any adverse effects associated with the supplementary intake of chromium at up to 1 mg/day. The dietary intake of trivalent chromium in European countries was also shown to be well below this amount. Similarly, the IOM did not set a UL, and the UK EVM also concluded that there is insufficient data from human or animal studies to derive an SUL for this nutrient. The EVM further noted that the oral toxicity of poorly absorbed trivalent chromium appears to be low, and it set a guidance level (GL) equivalent to 10 mg/person/day, which indicates that chromium III has a wide safety margin.

In the case of vitamin K, the SCF/EFSA could not establish a UL but the risk characterisation states that in the limited number of human studies carried out to date, there is no evidence of adverse effects associated with the supplementary intake of vitamin K in the form of phylloquinone at up to 10 mg/day for limited periods of time. The UK EVM derived a GL of 1 mg for daily supplementary intake and concluded that vitamin K (phylloquinone) would be unlikely to result in adverse effects in normal healthy individuals. However, because of the antagonistic interaction between phylloquinone and coumarin anticoagu-

Table 2. Qualitative risk characterisation for those vitamins and minerals for which no UL is available from the Scientific Committee on Food (SCF) and the European Food Safety Authority (EFSA)

Nutrient	Qualitative risk characterisation
Biotin	The SCF (2001) concluded that the risk of human toxicity from the usual dietary intake of biotin and from biotin supplements appears to be low. The SCF had insufficient data to draw any conclusions concerning the safety of very high level supplements. Although it was not possible to derive a numerical UL for biotin owing to lack of quantitative data, existing evidence from observational studies indicates that current levels of intake of biotin from all sources do not represent a health risk for the general population.
Chromium	No adverse effects have been convincingly associated with excess intake of chromium from food or food supplements. Overall, there is insufficient data from human or animal studies to derive a safe UL for chromium. However, the oral toxicity of the poorly absorbed trivalent chromium appears to be low. In a number of limited studies, there was no evidence of adverse effects associated with supplementary intake of chromium up to an intake of 10 mg chromium per day. The dietary intake of trivalent chromium in European countries is well below these amounts.
Pantothenic acid	Owing to the low toxicity of pantothenic acid and the lack of systematic oral dose-response intake studies, no LOAEL and NOAEL can be established, and hence no numerical UL can be derived. The SCF (2002) concluded from clinical studies using high doses of pantothenic acid that intakes considerably in excess of current levels of intake from all sources do not represent a health risk for the general population.
Riboflavin	No study has reported significant adverse effects in humans of excess riboflavin consumption from food or food supplements. Although this does not mean that there is no potential for adverse effects from high intakes, and it was not possible, based on the present database, to derive a UL, the limited evidence available from clinical studies indicates that current levels of intake of riboflavin from all sources do not represent a risk to human health.
Thiamin	Systematic data on adverse effects with oral intake of vitamin B_1 in humans are very limited. However, from the available literature it can be concluded that orally ingested vitamin B_1 has a very low risk of adverse effects. The US FNB (1998) concluded that no UL could be achieved if based on inadequate data. The SCF (1993) mentioned no evidence of toxicity at oral intakes up to 500 mg/day for one month. Previous evaluations of micronutrient safety classified vitamin B_1 as a nutrient with no adverse effects. The SCF (2001) concluded that it is not possible to derive a numerical UL for vitamin B_1. However, existing evidence from clinical studies as well as the long history of therapeutic use, at levels up to 200 mg/day for months, indicates that current levels of intake from vitamin B_1 from all sources do not represent a health risk for the general population.
Vitamin B_{12}	There are no adverse effects known for vitamin B_{12} from foods or from food supplements. There are no clearly defined adverse effects produced by vitamin B_{12} that can be used to define a LOAEL or NOAEL, which can be used as a basis for deriving a UL. The SCF (2000) concluded that there is no evidence that the current levels of intake from food and food supplements represent a health risk. Adverse effects have not been reported in the treatment of patients with compromised B_{12} absorption who received amounts up to 1000 μg/day orally for prolonged periods.
Vitamin K	Vitamin K_1 (phylloquinone), which is the form occurring naturally in food, is not associated with adverse effects in animal and human studies. A quantitative risk assessment cannot be carried out and a UL cannot be derived. In human studies with limited numbers of subjects, there is no evidence of adverse effects associated with supplementary intakes of vitamin K up to 10 mg/day for limited periods of time. However, people under medical supervision who were taking coumarin anticoagulant drugs should not increase their phylloquinone intake by dietary change or by using dietary supplements without medical advice because of their antagonistic interaction.
Vitamin C	Special case: see Section 5, group B nutrient.
Manganese	Special case: see Table 7, group C nutrient.

FNB, Food and Nutrition Board; LOAEL, lowest-observed-adverse-effect level; NOAEL, no-observed-adverse-effect level; UL, upper limit.

lant drugs, people taking these drugs should not increase their vitamin K intake significantly by dietary change or by using dietary supplements without medical advice. Similarly, vitamin K activity and clotting functions must be carefully considered before oral antibiotics are prescribed for patients taking coumarin drugs under medical supervision.

Potassium is another mineral for which the SCF/EFSA, IOM and EVM risk assessors were not able to establish an upper safe level. The SCF/EFSA concluded that the risk of adverse effects as a result of potassium intake from food sources (up to 5–6 g/day) is considered to be low for the general healthy population. Long-term intakes of about 3 g potassium per day as potassium chloride supplements, in addition to the intake from foods, have not been shown to cause adverse effects such as elevated plasma potassium or gastrointestinal symptoms in healthy adults. The EVM reported that supplementation studies have generally not revealed any side effects, and suggested that supplemental amounts of potassium up to 3.7 g/day appear to be without overt adverse effects. However, supplemental potassium in amounts of 5–7 g/day can cause adverse effects on heart function in apparently healthy adults, and elderly people may be more vulnerable due to reduced renal function. Overall, the clinical trial data on potassium chloride, together with the epidemiological evidence supporting the safety of larger amounts of potassium from fruits and vegetables, indicated that this nutrient has a wide margin of safety. Clinical trials showed no pattern of adverse effects for supplemental potassium at 1.5 g/day—a level that, if taken three times a day with meals with no greater than 0.5 g consumed on each occasion, could be considered to be the maximum safe level in food supplements (MSL).

Whereas potassium exerts a beneficial effect on hypertension by lowering blood pressure, the major adverse effect of sodium, taken as sodium chloride, is elevated blood pressure, which is a risk factor for ischaemic heart disease, stroke and renal disease. The habitual intake of sodium for populations across Europe is high and exceeds dietary recommendations. Risk assessors have not established a UL for sodium or sodium chloride. However, the EVM concluded that there appears to be a graded response to a range of intakes that included the current estimated intake levels, and that efforts should be directed at reducing salt in the diet, and therefore sodium chloride is not ordinarily suitable for use in food supplements.

4. Quantitative safety characteristics for nutrients with a UL

4.1 Establishment of a safe range of intake

The risk assessments for nutrients differ from other substances in foods because vitamins and minerals are essential for human life and consequently, adverse effects can result from deficiencies as well as from an excessive intake. In order to establish the risk to the population of exceeding the UL, the risk assessor has to draw together information on the range of safe intake—sometimes referred to as the 'acceptable range of oral intake (AROI; WHO 2002). For the purposes of the current risk management model the markers for the upper end of the safe range are the ULs, and for the lower end, the SCF reference labelling values (RLVs)—the recommended daily allowances (RDAs) for labelling purposes—are used (SCF 2003). Traditionally, RDAs are used as indicators to detect deficiencies for certain nutrients in the population, and some member states have proposed that the establishment of the maximum amounts should be based on, or limited to, multiples of the recommended daily amount (RDA) or population reference intakes (PRIs). However, this unscientific and arbitrary approach has been rejected by the EC and condemned by the Court of Justice (European Commission Orientation Paper 2007). It should be noted that the RDAs/PRIs and ULs are determined by two completely different conceptual approaches. However, these two values can be used as indicators to establish the extent of the range of safe intake and to help categorise nutrients on the basis of the risk associated with exceeding their ULs. In other words, where the UL and RDA/PRI are closer together, the safe range of intake is relatively small; where they are further apart, the safe range of intake is relatively large. The use of the reference labelling value (RLV) in the proposed model complies with the legal criterion that due account be taken of reference intakes of vitamins and minerals for the population, and helps to establish the breadth of the range of safe intakes, as well as a process for risk characterisation.

4.2 Intake data

4.2.1 Current intakes

Nutrient risk analysis requires an assessment of the potential intakes of vitamins and minerals from the various dietary sources. A fundamental problem is the adequacy of the information on nutrient intakes, and much greater attention needs to be placed scientifically on the acquisition, development and interpretation of exposure data for the essential nutrients for specific population groups (WHO 2002). Of particular relevance is the ongoing European Food Consumption validation (EFCOVAL) project (www.efcoval.eu), the aim of which is to further develop and validate a method to be used for the estimation of intake of foods, nutrients and potentially hazardous chemicals within the European population. In the absence of overall European Union (EU) data, national diet and nutrition surveys are the best sources of information, despite the fact that most of them have been conducted with different methodologies and may not be up to date. However, a pragmatic approach used in the current model, which has also been supported in the Commission's orientation paper, is to use the most appropriate available data. The UK National Diet and Nutrition Surveys (NDNS; UK Office for National Statistics 2003) and the North/South Ireland Food Consumption Survey (Irish Universities Nutrition Alliance (IUNA), 2001; Flynn, personal communication, 2007) provide the most recent and complete data available, and both reflect markets that are considered as 'mature' for both food supplements and fortified foods. In addition, specific nutrients can be assessed on a case-by-case basis using validated national data from individual EU member states.

For the purposes of the proposed model, it is necessary to determine whether current, actual or potential future intakes from food supplements and fortified foods are safe. Hence, the term 'mean highest intake' (MHI) from foods, including fortified foods, has been defined as the 97.5 percentile intake; the Irish data shown in Table 3 has been

Table 3. Micronutrient intakes in Irish adults aged 19–64 years (North/South Ireland Food Consumption Survey available at www.iuna.net, IUNA 2001; Flynn, personal communication 2007)

Nutrient	97.5 percentile (all sources)[a]		97.5 percentile (food only)[b]	
	Men	Women	Men	Women
Retinol, μg	2887	2885	2782	2521
Carotene, μg	6871	6842	6871	6735
Vitamin D, μg	13.5	15.1	11.2	9.6
Vitamin E, mg	28.3	38.3	14.5	12.5
Thiamin, mg	4.7	6.4	3.9	2.6
Riboflavin, mg	4.6	4.7	3.7	2.8
Vitamin B_6, mg	7.6	30.3[c]	5.9	3.8
Vitamin B_{12}, μg	14.8	15.1	13.2	11.6
Folic acid, μg	294	407	203	153
Preformed niacin, mg	52	44	47	32
Biotin, μg	92	103	75	54
Pantothenic acid, μg	12.5	14.4	10.5	7.3
Vitamin C, mg	588	588	209	187
Calcium, mg	1774	1419	1774	1296
Magnesium, mg	607	431	602	415
Phosphorus, mg	2710	1806	2710	1806
Iron, mg	28.8	72.2[c]	26.4	18.9
Zinc, mg	23.5	21.8	21.3	12.8
Copper, mg	3.5	3.0	3.3	2.8

[a] Food (including fortified foods) and food supplements.
[b] Food including conventional foods and fortified foods; excludes food supplements.
[c] The high values reflect the relatively small sample sizes when using P 97.5% values. The equivalent P 95% intake is 5.5 mg for vitamin B_6 and 25.5 mg for iron.

Table 4. Changes in nutrient intakes based on a comparison of National Diet and Nutrition Surveys (NDNS) undertaken in the UK in 1986/1987 and 2000/2001 (from UK Office for National Statistics 2003)

Vitamins	Mean intake for all men		
	1986/1987 UK adults survey	2000/2001 NDNS	% change in intake
A (μg)	1679	1017	−39%
Thiamin (mg)	2.01	2.22	+10%
Riboflavin (mg)	2.29	2.33	+2%
Niacin equivalents (mg)	40.0	46.4	+13%
B_6 (mg)	2.7	3.3	+22%
B_{12} (μg)	7.3	6.8	−7%
Folate (μg)	312	359	+15%
Pantothenic acid (mg)	6.6	7.8	+12%
Biotin (μg)	39	44	+13%
C (mg)	74.6	101.4	+36%
D (μg)	3.8	4.2	+11%
E (mg)	11.7	13.4	+15%
Minerals	**Mean intake for all men**		
	1986/1987 UK adults survey	2000/2001 NDNS	% change in intake
Iron (mg)	14	14	0
Calcium (mg)	940	1016	+8%
Potassium (mg)	3187	3371	+5%
Magnesium (mg)	323	311	−4%
Phosphorus (mg)	1452	1502	+3%
Copper (mg)	1.63	1.48	−9%
Zinc (mg)	11.4	10.7	−6%
Iodine (μg)	243	220	−9%

used in the proposed model for characterising risk, as shown in Section 5.

4.2.2 Future intakes

Future intake levels may change owing to changes in consumer preferences, food supplement use and the increased fortification of food products. Therefore, to gain a measure of potential changes to dietary patterns that might develop, a comparison of surveys undertaken in the UK in 1986–1987 and 2000–2001 was made (see Table 4; UK Office for National Statistics 2003). This shows the average adult intakes for men—the population group with the highest nutrient intakes generally—reflecting the higher energy intake compared with women. Although an increase in intake of more than 20% was recorded for only two of the vitamins (C and B_6), to take into account potential future changes in dietary patterns, the current model assumes a precautionary risk management factor of a 50% increase in dietary intake for all vitamins from foods, including fortified foods. In contrast to the data for vitamin consumption, the results for the UK over the 15 year period set out in Table 4 show a reduction in mean intake for men for many essential minerals, with only calcium intake increasing by more than 5%. Given that, for technical and taste reasons mineral fortification is self-limiting, the model assumes a precautionary risk management factor of 10% as the potential increase for mineral intakes. These precautionary risk management factors that allow for potential changes in intake from food, including fortified food, are used to estimate proposed maximum safe levels in food supplements (MSLs), as described in Section 6. There is also the potential for the intake of certain minerals via ingestion of water alone, water-based drinks and water used in the preparation of foods, and this amount has to be taken into account when assessing overall dietary intakes.

The intake data used in the model described in Section 5 are derived from the SCF/EFSA (2006) and EVM (2003). These intake levels can be modified to reflect existing and new data from other national and European surveys. Future risk scenarios will have to take into account the possibility of a lower dietary nutrient density if the trend for a reduced intake of certain minerals and vitamins continues in the future. The model can also be applied to different population groups, such as children, with the use of appropriate ULs, RDAs/PRIs and intake data.

5. Characterising risk using the population safety index (PSI)

The fundamental question is how to establish an objective process to determine how large the margin of safety is now and is likely to be in the future, allowing for varying dietary contexts, new research findings and their application to food products entering the food supply.

A concept of characterising risk using a PSI is set out in the following equation:

$$PSI = \frac{UL - (MHI + IW)}{RLV}$$

Where UL is the tolerable upper intake level, MHI is mean highest intake, IW is intake from water and RLV is reference labelling value. The calculated PSI values for adult men and women are shown in Appendix 1 and are based on the data from the North/South Ireland Food Consumption Survey, as shown in Table 3. These PSI values are similar to the values published by the European Responsible Nutrition Alliance (ERNA 2004), which utilised previous intake data that was based on the 97.5 percentile mean intake data from the Netherlands, Ireland, Italy and the UK for the highest risk group (male adults). The PSI conceptual approach provides a process by which the vitamins and minerals with ULs can be allocated into categories of risk. The proposed model assumes that where the PSI of a nutrient is higher than 1.5, i.e. where there is a margin of safety of 1.5 times the RLV or greater between the 97.5 percentile of intake from food (including fortified food), plus the estimated intake from water and the UL, the chance of exceeding the UL is low (group B). Where the PSI is 1.5 or below, i.e. the 97.5 percentile of intake from food (including fortified food) plus the estimated intake from water is either above the UL or less than 1.5 times the RLV below the UL, there is a potential risk of exceeding the UL (group C). In other words, the characterising factor for group B—'low risk of exceeding the UL'—is a PSI greater than 1.5 for adults, and for group C—'potential risk at excessive intakes'—a PSI of 1.5 or less. Group B also includes some nutrients as special cases: potassium, as described in Section 3, and vitamin C, folic acid and magnesium, as described below.

In the case of vitamin C, a UL was not established by the EFSA because of limited data. In contrast, the IOM found that the data was sufficiently conclusive to enable a UL of 2000 mg/day to be set. In view of the fact that mild gastrointestinal intolerance, e.g. flatulence and diarrhoea, are the most clearly defined adverse effects at high intakes of 3–4 g/day of vitamin C, it appears appropriate to use the IOM UL for calculating a PSI of 22.4. The UL established by the SCF/EFSA for synthetic folic acid is 1000 μg. Although there is no conclusive evidence in humans, the SCF/EFSA concluded that the risk of progression of the neurological symptoms in vitamin B_{12}-deficient patients as a result of folic acid supplementation cannot be excluded, and should be considered to be the most serious potential adverse effect. The data on the effect of intakes between 1 and 5 mg are limited, but in nearly all of the studies the neurological side effects of folic acid were associated with levels exceeding 5 mg/day. With the proposed increase in the RDA from 200 to 400 μg/day and the proposals to fortify bread and flour with folic acid, the potential effects on population intake will need to be monitored, especially in the elderly, who may be at risk with a marginal vitamin B_{12} nutritional status (Morris *et al.* 2007).

For magnesium, the amount in foods and beverages was not measured or taken into account in the calculation of the No Observed Adverse Effect Level (NOAEL) by risk assessors, and therefore the UL for magnesium cannot be derived for the intake from all sources. SCF/EFSA established a UL for magnesium of 250 mg/day for readily dissociable magnesium salts, such as those found in food supplements. As the UL of 250 mg/day is 67% of the SCF RLV, the decision to base the UL on magnesium salts means that a PSI cannot be calculated on the same basis as the other micronutrients. The proposed model develops a risk management process that reflects the relative risk associated with each group of nutrients, as shown in Table 5. Therefore, the concept of the PSI provides a tool by which risk categorisation can be achieved. The application of the PSI and the qualitative and quantitative information from the SCF/EFSA and other international risk assessments provides empirical and objective approaches to the classification of relative risk from the consumption of higher levels of vitamins and minerals. The categorisation of nutrients using the PSI approach is consistent with earlier classifications (Richardson 1997) and with the theoretical model for safe additions to foods that was prepared by the ILSI Europe Addition of Nutrients to Food Task Force (Flynn *et al.* 2003; Renwick *et al.* 2004).

6. Ensuring the micronutrient status and health of European populations

6.1 Micronutrient sufficiency

Whatever risk management approaches are adopted, the objective is to ensure the nutrient density of the diet of the population without jeopardising the safety of the food supply. The use of food supplements and the additions of nutrients to foods requires an understanding of national

Table 5. Characterisation of risk using quantitative population safety index (PSI) and qualitative assessments

GROUP A No UL set. No evidence of risk within ranges currently consumed; does not represent a risk to human health	GROUP B Low risk of exceeding the UL (PSI greater than 1.5)	GROUP C Potential risk of exceeding the UL or no UL set (PSI at or less than 1.5)
Vitamin B_1	Vitamin B_6	Vitamin A
Vitamin B_2	Vitamin C	β-carotene
Biotin	Vitamin D	Calcium
Vitamin B_{12}	Vitamin E	Copper
Pantothenic acid	Nicotinamide	Fluoride
Vitamin K	Molybdenum	Iodine
Chromium (trivalent)	Phosphorus	Iron
	Selenium	Manganese
	Magnesium	Zinc
	Folic acid	
	Potassium	

dietary patterns and the nutritional status of the population or target groups. The use of diversified diets, foods with added nutrients and food supplements can undoubtedly improve the health of those segments of the population that are vulnerable and susceptible to deficiencies. Although there is a need for caution—especially in such dynamic areas of nutrition research and because of changing dietary habits and patterns—in Europe, there is clearly a widening gap between the ideal, balanced diet and the reality of what people eat. Poor food choices, excessive intakes of calories and fat, low intakes of fruits and vegetables and weight-reducing dietary regimes are now impacting on the nutritional status of substantial numbers of people in Europe (Eurodiet 2001; Godfrey and Richardson 2002). Although average intakes of most nutrients in the diets of adults and children may be greater than or equal to the RDA, these average figures fail to show that intakes of several vitamins and minerals for significant proportions of the population fall well below the RDA, reference nutrient intake (RNI) or PRI, and in some cases, the lower reference nutrient intake (LRNI). Furthermore, there are sizeable groups of the population that are at significant risk of deficiency. For example, iron deficiency is one of the main nutritional disorders in Europe (Hallberg 1995). A high prevalence of depleted iron stores and even overt iron deficiency anaemia is observed in women of childbearing age and in children (Hallberg and Asp 1996). Other nutrients of concern in Europe include iodine, folate and vitamin D (Sandstrom 2001).

In the UK, the NDNS for children (4–18 years), adults (19–64 years) and for those over 65 years of age (UK Office for National Statistics 1998, 2000, 2003) need careful analysis to show not only the most up to date assessment of the nutritional status of the population, but also to check if the situation in the UK is improving or not. The results of the most recent NDNS in 2003 for adults aged 19–64 years are particularly concerning because of failures of specific groups to achieve either the RNI or LRNI (UK Office for National Statistics 2003). For example, for folic acid, 86% of women aged 35–49 years had an intake from all sources (including supplements) of less than 400 μg/day. This low intake was in spite of the Department of Health advice to women of childbearing age to take a daily supplement containing 400 μg of folic acid per day in addition to the 200 μg obtained from the diet. For iron, 91% of women overall and 96% of women aged 19–24 years, 41% aged 25–34 years and 27% aged 35–49 years, had intakes below the LRNI. For magnesium, 74% of women overall and 85% of women aged 19–24 years failed to achieve the RNI. Overall, one in ten men and one in seven women had intakes below the LRNI for magnesium. More importantly, as previously shown in Table 4, the intakes of vitamin A, B_{12}, iron, magnesium, zinc, copper and iodine have fallen since the previous NDNS survey in 1986/1987. For the elderly, the proportions of the population with marginal and overt deficiencies are even worse (Richardson 2007). The Food Standards Agency (2007) published a survey of a national representative sample of the most materially deprived, low-income households in the UK. Analysis of population groups, broken down by age and sex as well as region, household type and ethnic group, highlighted significant proportions of respondents with intakes below the LNRI for several essential vitamins (e.g. vitamin A, folate and riboflavin) and minerals (e.g. calcium, magnesium, iron, iodine and zinc). These results demonstrate the increasing likelihood of dietary inadequacy in the most vulnerable low-income groups.

6.2 Balancing risks of excessive intakes and deficiency

Where there is a narrow range of safe intake and a potential risk that consumers may exceed the UL on a daily basis, it may not be possible for the risk manager to use the UL or guidance level (GL) as a reference point for establishing a maximum level for a particular nutrient. A good example is the preformed retinol form of vitamin A. The SCF/EFSA and US IOM established a UL of 3000 μg retinol equivalents (RE) per day, which was based on the risk of adverse effects including birth defects, reduced bone mineral density and liver abnormalities. Above this level the risk of adverse effects gradually increases and long-term vitamin toxicity is associated with levels of 7500–15 000 μg preformed retinol per day. In 2003, the UK EVM considered the two areas of evidence regarding potential adverse effects of vitamin A, one concerning the risk of teratogenicity and the other the risk of bone fracture. The EVM concluded that, given the severity of the effect, it is prudent to regard 3000 µg RE/day as the threshold for teratogenicity. However, in setting the GL, the EVM also took into account emerging evidence from epidemiological studies and from studies in laboratory animals that vitamin A affects calcium metabolism and may increase bone fracture risk in older women and men. Despite the paucity of data, the EVM concluded that the risk of hip fracture is a continuous graded response, that it was not possible to identify an intake that is without some degree of risk, and that the GL should be 1500 µg RE/day. In comparison, the RLV for vitamin A is 800 µg/day.

The example of preformed retinol illustrates that the determination of a safe level of intake is even more critical when the range of intakes for different age and sex groups is very wide, reflecting the limited distribution of preformed retinol in foods. This dependency on retinol from relatively few foods (the main dietary sources are liver and liver products) results in the median intakes typically being 20–50% less than the average intakes, which highlights the risk of inadequate intakes in sizeable groups of the UK population. Whereas the dietary intake data shows that the risk of exceeding the UL could theoretically affect a very small proportion of the population (less than 3%), there are substantial groups of the population that fail to achieve the RDA and thus have a real risk of deficiency. Figure 1 illustrates the intake of preformed retinol from all sources, including food supplements in the UK and the proportions of the population at risk of either deficiency or excess (ERNA 2004).

Epidemiological and clinical data that has accumulated over the past 10 years indicates that appropriate intakes of vitamin D may provide greater benefits than previously thought (Giovannuci 2006; Vieth *et al.* 2007). This suggests a need for levels of supplementation or food fortification, or both, that are higher than currently employed. The potential of vitamin D to produce adverse effects may also have been overcautious, and the newer clinical trial data shows that vitamin D is unlikely to cause adverse effects at intakes higher than were previously considered unsafe (Vieth 2006; Hathcock *et al.* 2007). The absence of adverse effects in trials conducted with healthy adults suggests a UL for vitamin D of 250 µg/day, which would be obtainable from all sources (Hathcock *et al.* 2007). The conclusions from several researchers suggest that the present ULs for vitamin D established by the IOM and SCF/EFSA are lower than those justified by the recent scientific evidence, and are hence in urgent need of review.

Using these illustrations for vitamin A and vitamin D, risk managers have to address the overall balance of the diet, potential nutrient interactions and the contribution of specific foods and food supplements to nutrient intakes, in addition to new research findings and the use of clear,

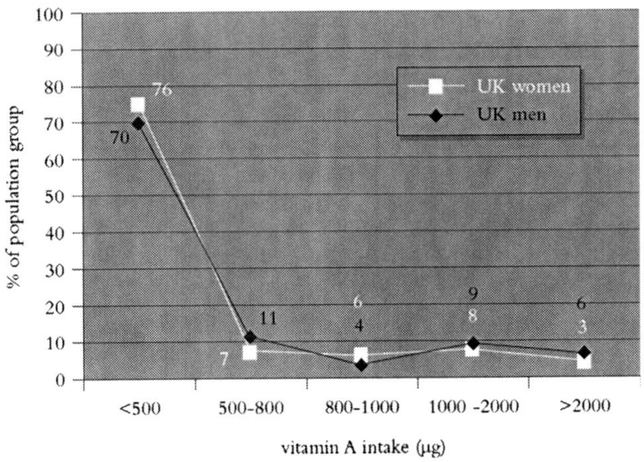

Figure 1. Intake of preformed vitamin A from all sources in the UK.

understandable labelling. In addition, risk managers must take into account, on a case-by-case basis, the risks of both deficiency and excessive intakes, particularly for the group C nutrients.

7. Setting maximum safe levels in food supplements (MSLs)

In summary, taking into consideration the risk categorisation of nutrients and the quantitative estimates of potential higher intakes from all other food sources including fortified foods, the proposed risk management model has been applied to determine maximum levels of vitamins and minerals in food supplements, as set out in Table 6. The MSLs per day for group B have been calculated as follows:

$$\text{For vitamins: MSL} = \text{UL} - (\text{MHI} \times 150\%)$$

$$\text{For minerals: MSL} = \text{UL} - [(\text{MHI} \times 110\%) + \text{IW}]$$

Intake data from different countries could lead to some variations in MSLs for the individual nutrients. The MSLs in Table 6 that were derived using the data from the North/South Ireland Food Consumption Survey in Table 3 are very similar to the values already set out by the European Responsible Nutrition Alliance (ERNA 2004), and which are included in the EC's orientation paper (2007). Hence, for consistency the proposed MSLs per day, which are shown in Table 6, are those in the ERNA report. Table 6 also contains the values used as industry guidelines that were based on risk assessments of the individual nutrients carried out over 10 years ago by Shrimpton (1997).

The case-by-case qualitative risk management assessments of group C nutrients are shown in Table 7. In the case of group C micronutrients, for which there is a potential risk of exceeding their ULs, all the nutrients have a PSI of less than 1.5. All of these micronutrients have

Table 6. New proposed maximum safe levels in food supplements (MSLs) together with current food supplement industry guidelines for MSLs per day, as developed in 1997

	New proposed MSL/day	Current food supplement industry guidelines for MSLs/day (1997)[a]
GROUP A		
Vitamin B_1 (mg)	—	100
Vitamin B_2 (mg)	—	200
Biotin (μg)	—	500
Vitamin B_{12} (μg)	—	500
Pantothenic acid (μg)	—	500
Vitamin K (μg)	—	—
Chromium (mg)	—	0.2
GROUP B		
Vitamin B_6 (mg)	18–93[b]	200
Vitamin C (mg)	1650	2000
Vitamin D (μg)	35	10
Vitamin E (mg)	270–970[b]	800
Nicotinamide (mg)	820	450
Molybdenum (μg)	350	200
Phosphorus (mg)	1250	1500
Selenium (μg)	200	200
Magnesium (mg)	250	350
Folic acid (μg)	600[c]	400
Potassium (mg)	1500	—
GROUP C[d]		
Vitamin A (μg)	800–1000	2300
β-carotene (mg)	4.8–7	20
Calcium (mg)	1000–1500	1500
Copper (mg)	1–2	5
Fluoride (mg)	3.5	—
Iodine (μg)	150–200	500
Iron (mg)	14–20	15
Manganese (mg)	2	15
Zinc (mg)	10–15	15

[a] Data taken from Shrimpton 1997; Shrimpton and Richardson 2003.
[b] In the case of vitamins B_6 and E, the risk assessments undertaken by the SCF/EFSA and the US IOM have produced widely divergent ULs. Taking into account these discrepancies, the MSL could ultimately vary between 18 and 93 mg/day for vitamin B_6 and 270 and 970 mg/day for vitamin E.
[c] The UL established by the SCF/EFSA of 1000 μg/day for folic acid is for supplementation only. Folic acid is found foremost in breakfast cereals at levels around 100–150 μg per serving. Under current levels of fortification, an individual is unlikely to exceed 400 μg/day. An MSL up to 600 μg/day would therefore currently be appropriate. An increase of the RDA from 200 to 400 μg/day and the moves towards folic acid fortification of breads and flours will necessitate careful monitoring of intake levels within the population.
[d] For the case-by-case risk management of group C nutrients and proposed MSLs see Table 7.

recognised adverse effects; where there is a potential risk that consumers may exceed the UL on a daily basis, it may not be possible for risk management to establish a meaningful MSL in relation to the UL. For micronutrients in this category such as vitamin A (preformed retinol), calcium, copper and zinc, the risk management process needs to take account not only of the risk of excessive intake but also the risk of insufficient intake. In seeking a balance between risk of deficiency and risk of excess, a pragmatic risk management approach has been adopted to reflect concerns on low current intakes of these essential nutrients in certain population groups.

The proposed risk management model set out in this paper has therefore been used as an aid to the setting of MSLs per day as required by the European Food Supplements Directive.

8. Setting maximum amounts of vitamins and minerals in fortified foods

The most comprehensive theoretical model has been prepared by the ILSI Europe Addition of Nutrients to Foods Task Force (Flynn *et al.* 2003). The methodology identifies a number of critical factors that determine the risk of unacceptably high intake for each micronutrient at high levels of food/energy intakes. These factors include:

- ULs
- High micronutrient intakes in Europe at the 95th percentile intake for each nutrient
- The proportion of foods to which micronutrients could practically be added
- A range of estimates for the fractions of foods that might actually be fortified for each nutrient.

A maximum level was set up for each micronutrient per typical serving or 100 kcal portion. Reassuringly, the ILSI model also resulted in three categories of risk characterisation for the micronutrients that could be added safely to appropriate foods at specified levels. In addition, preformed retinol was highlighted as a special case. The ILSI model has been modified to consider the intakes from food supplements and to take into account the impact of the new regulation on nutrition and health claims (European Parliament and of the Council 2007). More recently, Rasmussen *et al.* (2005) and Kloosterman *et al.* (2007) have also proposed risk assessment approaches that are based on the ILSI model, all of which express the maximum levels of certain nutrients in fortified foods in weight units per 100 kcal.

Whilst for food supplements the maximum amount will be set per daily amount of consumption of the supplement (e.g. per one/two/three tablets or capsules, or other measure indicated by the manufacturer), for fortified foods the maximum level is likely to be set per weight or per energy of the food consumed (e.g. per 100 kcal, per 100 g/ml, per portion). The EC's orientation paper (2007) states that, in order to ensure that health-driven practices are preserved and a minimum significant amount of the nutrient is present in foods without a risk of excess intakes,

Table 7. Case-by-case risk management assessment of group C nutrients

Nutrient	Qualitative risk characterisation
Vitamin A (preformed retinol)	The teratogenic effect of preformed retinol on the newborn child is well documented because of the severe and irreversible nature of this form of toxicity. The UL applies to both dietary and supplemental intakes of vitamin A. The 97.5 percentile intake for adults in most of Europe is greater than 3000 μg RE/day. As the RDA for vitamin A is 800 μg RE/day and the distribution of intakes is great, especially in relation to consumption of liver and liver products, risk management for preformed retinol is particularly challenging to avoid deficiency as well as excess. Recent epidemiological data has indicated that the risk of hip fracture in older people may be associated with intakes as low as 1500 μg RE/day. SCF/EFSA (2006) states that, because current intakes may exceed the UL, careful consideration should be given to the appropriateness of the enrichment of human foods with vitamin A, and to the potential effects on human exposure of the addition of vitamin A to animal feed.
β-carotene	Until recently, β-carotene was considered to show no adverse effects in human. However, there may be a very small difference between the levels that may confer health benefits (up to 10 mg/day, mainly from natural sources) and those that may produce adverse effects in smokers in the general population (20 mg/day). The promise of β-carotene for cancer prevention led to two large-scale primary prevention studies in very high risk populations of chronic heavy smokers. β-Carotene supplementation at levels of 20 mg/day resulted in an increased incidence of lung cancer in the high risk groups. The SCF/EFSA state that, in this situation, it seems that the use of β-carotene as a supplement should be regarded cautiously.
Calcium	The evidence supporting an increased risk of kidney stones with a high calcium intake is far from clear and it has been contradicted by more recent credible evidence (IOM 1997). High dietary calcium levels can influence the bioavailability and absorption of many trace elements in individuals with low intakes, e.g. magnesium, iron, manganese and zinc. Acute adverse effects relate to constipation, abdominal pain and diarrhoea. Subpopulations known to be susceptible to high levels of calcium include individuals with renal insufficiency, alkalosis and dehydration due to vomiting and anorexia, and those using thiazide diuretics. The SCF/EFSA and IOM both derived a UL of 2500 mg/day for total intake from all sources and the EVM set a GL for supplemental calcium at 1500 mg/day, stating that such a supplemental level would not be expected to result in any adverse effect.
Copper	The occurrence of either acute or chronic copper toxicity in humans is rare and tends to be confined to certain populations with high copper concentrations in drinking water and to those individuals who have a hereditary predisposition to copper toxicity (Wilson's disease). Liver damage is used as a reliable indicator of the long-term ingestion of copper, and the NOAEL of 10 mg/day is based on the absence of any adverse effects on liver function using supplemental copper gluconate. From essentially the same studies, IOM and SCF/EFSA derived ULs of 10 mg and 5 mg/day, respectively, and the EVM derived an SUL from animal studies of 10 mg for supplemental intakes. From the available data, copper can cause adverse effects in humans. However, as there appears to be an absence of adverse effects at intakes in the range of 10–12 mg/day, and typical and 97.5 percentile intakes are less than 2 and 3 mg/day, respectively, a supplemental intake of 1–2 mg is acceptable. Acute copper toxicity in drinking water appears to have a threshold of approximately 6 mg/L. The SCF/EFSA notes that copper intakes from drinking water may be appreciable, and may thus need to be taken into account.
Fluoride	Fluoride toxicity is well known and has been extensively reviewed. The critical adverse effects are dental fluorosis in children and skeletal fluorosis in adults. Excessive intake in children manifests itself as mottled brown discolouration of the teeth. Skeletal fluorosis carries an increase risk of bone fracture. Fluoridation of drinking water is often carried out as a public health measure. Skeletal fluorosis is associated with intakes of 6 mg of fluoride or more per day from this source. The SCF/EFSA set a UL of 7 mg/day for adults and 5, 2.5 and 1.5 mg/day for children aged 9–14, 4–8 and 1–3 years, respectively. Typical adult dietary intakes are around 3–4 mg/day. In establishing a maximum level, intake from food, water and dental products must be considered.
Iodine	Excessive intake of iodine can occur as a result of the ingestion of large amounts of seaweed, kelp, marine fish, ground beef containing thyroid, iodised water, bread or salt and iodide-containing food supplements. Toxic effects are not observed in humans until daily intakes have exceeded 11 000 μg, but intakes of 2000 μg are regarded as excessive and potentially harmful. Except for rare cases of hypersensitivity, humans can tolerate high intakes of iodine because of biological mechanisms that protect against such exposure. A change in thyroid function with elevated levels of thyroid stimulating hormone (TSH) is used as an indicator of an increased risk of developing clinical hypothyroidism and is the critical adverse effect on which to base the UL. The SCF/EFSA set a UL of 600 μg/day, and data from European populations indicates that intakes of iodine from all sources in adults are unlikely to exceed the UL. For example, for the UK, where iodine intake is considered to be high relative to other European countries, the 97.5 percentile intake in men is 434 μg.

(*Continued*)

Table 7. (*Continued*)

Nutrient	Qualitative risk characterisation
Iron	Adverse gastrointestinal effects (i.e. nausea, epigastric discomfort and constipation) have been reported after short-term oral dosage at 50–60 mg/day with supplemental non-haem iron preparations, particularly if taken without food. Elevated serum ferritin levels and transferrin saturation are indicators of iron overload. Epidemiological associations between high iron intake and/or stores and an increased risk of chronic diseases such as cardiovascular disease, type 2 diabetes and cancer are conflicting and do not provide convincing evidence of a causal relationship. Based on current estimates of iron intakes in European countries, the risk of adverse effects from high iron intake from food sources, including fortified foods but excluding food supplements, is considered to be low. Intake from food supplements in men and postmenopausal women may increase the proportion of the population that is likely to develop biochemical indicators of high iron stores. Some at-risk groups with poor iron status, such as menstruating women and children, could benefit from supplementary iron intakes. Up to 0.5% of the population is sensitive to iron overload (hereditary haemochromatosis) and should avoid iron supplements and highly iron-fortified foods. The SCF/EFSA could not establish a UL. The IOM identified a supplemental LOAEL of 60 mg/day, to which it added 10–11 mg/day from dietary sources, and subsequently derived a UL from all sources of 45 mg/day for adults. The EVM derived a GL of 17 mg/day for supplemental iron.
Manganese	Miners and smelters that are chronically exposed to manganese dust suffer from 'manganism'—a neurotoxic condition similar to Parkinson's disease. Oral intake of manganese via drinking water is associated with neurological and behavioural effects in the elderly. Owing to limitations in the human data, the SCF/EFSA could not establish a UL. The IOM found no evidence of toxicity at intakes of less than 11 mg/day and set a UL of 11 mg/day. The EVM established a GL of 4 mg/day based on data that indicated no adverse effects from 4 mg of manganese in addition to the amounts present in foods (mean intake: 4.9 mg; 97.5 percentile: 8.2 mg/day) and a GL of 0.5 mg/day for 'older' people. The risk of an adverse effect from excess manganese from food and food supplements appears to be low. At the same time, the safety margin in both humans and animals also appears to be low.
Zinc	Excessive intake can cause adverse effects in humans and animals. In humans, the effects of acute zinc toxicity are gastrointestinal disturbances giving rise to abdominal pain, nausea and vomiting. Chronic zinc toxicity is associated with changes in copper balance leading to symptoms of copper deficiency. The SCF/EFSA noted an absence of any adverse effects on a wide range of indicators of copper status at an intake of 50 mg/day (NOAEL), and recommended a UL of 25 mg/day. The IOM UL for zinc is 40 mg/day from all sources, and the EVM derived an SUL of 25 mg for supplemental zinc. The available studies show that mean zinc intakes of adults and children in EU countries are below the UL. The 97.5 percentiles of total zinc intakes for all age groups are close to the ULs, which, in the view of the SCF/EFSA, are not a matter of concern.

EU, European Union; EFSA, European Food Safety Authority; EVM, Expert Group on Vitamins and Minerals; GL, guidance level; IOM, Institute of Medicine; LOAEL, lowest-observed-adverse-effect level; NOAEL, no-observed-adverse-effect level; RE, retinol equivalents; RDA, recommended daily amount; SCF, Scientific Committee on Food; SUL, safe upper limit; UL, upper limit.

there may be a need for limitations and restrictions to the addition of certain vitamins and minerals in normal foods (e.g. addition of iodine restricted only to salt and addition of vitamins A and D reserved for specific products).

Typically, the amounts of nutrients added to food products are based on making a nutrient content claim. The new Regulation on Nutrition and Health Claims (2007) sets out the criteria for 'source' and 'high' in vitamins and minerals. A claim that a food is a source of vitamins and/or minerals, and any claim likely to have the same meaning for the consumer, may only be made when the product contains the vitamins and/or minerals in at least a significant amount, as defined in the Annex of Directive 90/496/EEC; i.e. 15% RDA per 100 g or 100 ml of a food, or per package of a food if the package contains only a single portion. A claim that a food is 'high' in vitamins and/or minerals may only be made if the product contains at least twice the value of 'source'. In the case of foods that are a natural source or are high in vitamins and minerals, the term 'naturally' may be used as a prefix to the claim.

As a result of the legislation, most nutrients are added to foods to deliver either 15 or 30% RDA per 100 g, per 100 ml or per quantified serving, unless there is a specific nutritional rationale to have higher proportions of the RDA. For the most common fortificants, which are in group A and group B of the risk characterisation (e.g. vitamin B_1 (thiamin), vitamin B_2 (riboflavin), nicotinamide, folic acid, vitamin B_6, vitamin B_{12}, biotin, pantothenic acid, vitamin D and vitamin E), the margins for additional fortification above current levels that are already accounted for in the current risk management model are sufficiently large to pose a very small risk to human health from excessive intakes. Some nutrients used in food supplements tend not to be added to food products either for technical reasons or because consumers have not been made aware of their particular nutritional benefits, e.g.

chromium, vitamin K and several trace elements such as magnesium, phosphorus, manganese, zinc and copper. β-Carotene is used as a food additive or as provitamin A. Iodine, as previously noted, is commonly added to table salt. Attempting to add nutrients such as iron, calcium and magnesium at high levels often presents technological problems with colour, texture and taste, as well as having implications for the shelf-life of products (Richardson 1997). Folic acid fortification of cereal grain products is aimed at preventing neural tube defects. However, concerns have been expressed over the possibility that an increased folic acid intake might delay the diagnosis of vitamin B_{12} deficiency and even exacerbate its neurological and neuropsychiatric effects in the elderly. The policy decisions and scientific issues related to folic acid fortification have been reviewed by Smith (2007).

Bearing in mind the scientific discussions and technical limitations, as well as the fact that the consumption of nutrient-dense and fortified foods is constrained by the energy density and satiating aspect of the food or meal, the risk of excessive intakes of vitamins and minerals from foods is small. Preformed retinol merits further detailed consideration, as noted previously. Nevertheless, present levels of retinol addition to foods, e.g. restoration, substitution and fortification, as well as current levels used in food supplements appear to pose no problems in the diets of Europeans. The impact of fortified foods on total dietary consumption was analysed by Godfrey *et al.* (2004). The percentage of each food category that was fortified was estimated. The data indicated that, overall, about 75% of foods and drinks consumed in European diets are rarely or never fortified. Fortified foods were found to rarely contribute more than 3% of the total diet on a per capita basis, an exception being in countries where it is mandatory to fortify a staple food (e.g. as in the case of flour in the UK). The authors estimated that high-level consumers of fortified foods are unlikely to obtain more than 10% of their diet in fortified form.

Recently, the Confederation of Food and Drink Industries in the EU (CIAA, personal communication 2007) has proposed a basis for setting maximum levels for vitamins and minerals in food using the same groups (A, B and C) to characterise risk, as shown in Table 5. The methodology is currently being developed to decide the best way to set the maximum amounts of the nutrients, i.e. per weight or per energy of the food consumed (e.g. per 100 kcal, per 100 g/ml, per portion).

9. Conclusion

The risk assessments for determining ULs depend on the availability of good data on the nature, frequency and severity of adverse effects detected at different levels of intake. The database supporting the safety-in-use of vitamins and minerals is limited and there is rarely adequate consideration given to potentially vulnerable population subgroups such as children or the elderly. It is important for risk managers to recognise that the UL is defined as the maximum level of *chronic* daily intake of a nutrient (from all sources) that is judged to have no appreciable risk of an adverse effect occurring at some specified level of exposure. Similarly, the UK EVM defined safe upper levels (SULs) as an intake that can be consumed daily over a lifetime without any significant risk to health, on the basis of available evidence. GLs are also defined by the EVM as levels that represent an approximate indication of levels that would not be expected to cause adverse effects but have been derived from limited data and are less secure than SULs. The determination of SULs or GLs relate to the amounts of vitamins and minerals that potentially susceptible individuals could take daily on a lifelong basis without medical supervision and in reasonable safety (EVM 2003). The setting of upper safe levels by risk assessors and the setting of maximum levels in foods and food supplements by risk managers builds in levels of precaution and provides a framework within which the consumer can make an informed decision about intake, having confidence that harm should not ensue. The upper safe levels set in risk assessments tend to be conservative, and it is possible that for some vitamins and minerals, e.g. vitamin B_6 and vitamin C, larger amounts could be considered for shorter-term consumption because the available data are limited and relate to differing time periods. As a consequence of the limited data, risk assessors build in uncertainty factors and apply the precautionary principle of allowing for the variable quality of information and for risk managers to weigh up the different safety margins between necessity and adverse effects. As the market, and hence the exposure, varies over time, risk managers will need to monitor changing patterns of addition of nutrients to different foods and the use of food supplements in order to review the risk management options. There will also be a need for regular review of the scientific basis for the establishment of lower and upper safe intakes of micronutrients, particularly for those nutrients with narrow safe ranges of intake. Any changes in either RDAs or ULs will also impact on the risk management options. There are broadly two kinds of food risks to the public that relate to food supplements or food fortification (May 2000). Where the risks are voluntary, such as taking a food supplement or buying a fortified food, the role of the risk manager is to ensure that the underpinning regulations are in place and to ensure that individuals are given sufficient information and advice to enable them to make their own choices, with the assurance that the regulatory environment exists to limit the risk. In the second kind of food risk, there may be a wider or more serious public health concern when an individual may have no meaningful choice (e.g. mandatory addition of nutrients to flour and margarine in the UK), or when vulnerable groups such as children or the elderly are involved. Therefore, risk managers need to examine a range of options, taking into account known usage patterns, setting out clearly the pros and cons and assessing the overall implications of their decisions. Clearly, with respect to the essential nutrients, there are two types of risk: those of deficiency and those associated with excessive intake, leading to adverse effects. Food fortification practices and current levels of nutrients used in food supplements for over two decades have been shown to be safe and effective. The objective of the model for risk management set out in this paper is to assist in the development and implementation of an approach to set maximum levels in fortified food and food supplements that are consistent with the latest scientific assessments, with good sustainable industry practices, and with efforts to enhance consumer understanding. The applicability of the model described in this paper depends on the availability of the input data, particularly data on nutrient intakes from food consumption surveys that cover all member states of the EU. Currently, such data are not available. However, the PSI calculation has now been applied to several different sets of intake data (ERNA 2004), and in the current paper the method has been applied using the most comprehensive data from Ireland. It is proposed that the approaches used in the current risk management model and the MSLs derived from it could be applied in all member states. It is the purpose of this review to contribute towards the development of an appropriate process for the setting of maximum levels in fortified foods and food supplements in the EU. Consultation and dialogue between the various stakeholders are critical to ensure that proportionate measures are used to protect the consumer and to facilitate informed choice.

10. Acknowledgements

The production of this paper was supported by the European Responsible Nutrition Alliance (ERNA), the UK Council for Responsible Nutrition (CRN) and the European Federation of Associations of Health Product Manufacturers (EHPM). The author also acknowledges editorial comments from Astrid Kruizinga (TNO Quality of Life, The Netherlands), Peter Berry Ottaway, Sam Jennings (Berry Ottaway Associates Ltd, UK), David Godfrey (Nutragen Ltd, UK) and Patrick Coppens (ERNA, Belgium).

11. References

Codex Alimentarius Commission. 2003. *Joint FAO/WHO Food Standards Programme. Proposed draft guidelines for vitamin and mineral food supplements.* ALINORM 04/27/26. Bonn, Germany. Available at: www.codexalimentarius.net

Confederations des Industries Agro-alimentaries de l'UE. 2007. *Proposed CIAA position for setting maximum levels of vitamins and minerals in food.* CIAA, Brussels, Belgium. Available at: www.ciaa.be

Council of the European Communities. 1990. Council Directive 90/496/EEC of 24th September 1990 on nutrition labelling for foodstuffs. *Official Journal of the European Communities* **L276**: 40-44.

Eurodiet. 2001. Nutrition and diet for healthy lifestyles in Europe: the EURODIET evidence. *Public Health Nutrition* **4**: 2A-2B.

European Commission Scientific Committee on Food. 2003. *Opinion of the Scientific Committee on Food on the revision of reference values for nutrition labelling.* SCF/CS/NUT/GEN/18 Final. Available at: http://ec.europa.eu/food/fs/sc/scf/out171_en.pdf

European Commission. 2007. *Orientation paper on setting maximum and minimum amounts for vitamins and minerals in foodstuffs.* SANCO/E4/FDA/bs D/540510, Brussels, Belgium.

European Parliament and of the Council. 2002. Directive 2002/46/EC of 10th June 2002 on the approximation of the laws of the member states relating to food supplements. *Official Journal of the European Communities L183/51-L183/57.*

European Parliament and of the Council. 2006. Regulation (EC) No. 1925/2006 on the addition of vitamins and minerals and of certain other substances to foods. *Official Journal of the European Union L404/26-L404/38.*

European Parliament and of the Council. 2007. Corrigendum to Regulation (EC) No. 1924/2006 of 20th December 2006 on nutrition and health claims made on foods. *Official Journal of the European Union L12/3-L12/18.*

European Responsible Nutrition Alliance. 2004. *Vitamin and mineral supplements: a risk management model.* ERNA, Brussels, Belgium. Available at: www.erna.org

Flynn, A., Moreiras, O., Stehle, P., Fletcher, R., Muller, D. and Rolland, V. 2003. Vitamins and minerals: a model for safe addition to foods. *European Journal of Nutrition* **42**: 118-130.

Food and Agriculture Organisation and International Life Sciences Institute. 1997. *Preventing micronutrient malnutrition: a guide to food-based approaches. A manual for policy makers and progressive planners.* ILSI Press, Washington DC, USA.

Food Standards Agency. 2007. Low income diet and nutrition survey. Available at www.food.gov.uk/science/dietarysurveys

Giovannucci, E. 2006. The epidemiology of vitamin D and colorectal cancer: recent findings. *Current Opinion in Gastroenterology* **22**: 24-29.

Godfrey, D. and Richardson, D.P. 2002. Vitamins and minerals for health. *British Food Journal* **104**(11): 913-933.

Godfrey, D., Tennant, D. and Davidson, J. 2004. The impact of fortified foods on total dietary consumption in Europe. *British Nutrition Foundation Bulletin* **29**: 188-198.

Hallberg, L. 1995. Results of surveys to assess iron status in Europe. *Nutrition Reviews* **11**: 314-322.

Hallberg, L. and Asp, N.G. 1996. *Iron nutrition in health and disease*: 364. John Libbey, London, UK.

Hathcock, J.N., Shao, A., Vieth, R. and Heany, R. 2007. Risk assessment for vitamin D. *American Journal of Clinical Nutrition* **85**: 6-18.

Institute of Medicine (IOM). 1999. *Food and Nutrition Board. Dietary reference intakes for calcium, phosphorus, magnesium, vitamin D and fluoride.* National Academy Press, Washington DC, USA.

Institute of Medicine (IOM). 2000a. *Food and Nutrition Board. Dietary reference intakes for thiamin, riboflavin, niacin, vitamin B_6, folate, vitamin B_{12}, pantothenic acid, biotin and choline.* National Academy Press, Washington DC, USA.

Institute of Medicine (IOM). 2000b. *Food and Nutrition Board. Dietary reference intakes for vitamin C, vitamin E, selenium and carotenoids.* National Academy Press, Washington DC, USA.

Institute of Medicine (IOM). 2002. *Food and Nutrition Board. Dietary reference intakes for vitamin A, vitamin K, arsenic, boron, chromium, copper, iodine, iron, manganese, molybdenum, nickel, silicon, vanadium and zinc.* National Academy Press, Washington DC, USA.

Irish Universities Nutrition Alliance. 2001. *North/South Ireland Food Consumption Survey.* Available at: www.iuna.net

Kloosterman, J., Fransen, H.P., de Stoppelaar, J., Verhagen, H. and Rompelberg, C. 2007. Safe addition of vitamins and minerals to foods: setting maximum levels for fortification in The Netherlands. *European Journal of Nutrition* **46**(4): 220-229.

May, R. 2000. *Review of risk procedures used by the Government's advisory committees dealing with food safety.* Report of a group led by Sir Robert May, Chief Scientific Adviser. Department of Trade and Industry, UK.

Micronutrient Initiative. 2007. Available at: www.micronutrient.org

Morris, M.S., Jacques, P.F., Rosenberg, I.H. and Selhub, J. 2007. Folate and vitamin B_{12} status in relation to anaemia, macrocystosis and cognitive impairment in older Americans in the age of folic acid fortification. *American Journal of Clinical Nutrition* **85**(1): 193-200.

Morris, M.S., Jacques, P.F., Rosenberg, I.H. and Selhub, J. 2007. Folate and vitamin B_{12} status in relation to anaemia, macrocystosis and cognitive impairment in older Americans in the age of folic acid fortification. *American Journal of Clinical Nutrition* **85**(1): 193-200.

Rasmussen, S.E., Andersen, N.L., Dragsted, L.O. and Larsen, J.C. 2005. A safe strategy for addition of vitamins and minerals to food. *European Journal of Nutrition* **45**(3): 123-135.

Renwick, A., Flynn, A., Fletcher, R., Muller, D., Tuijtelaars, S. and Verhagen, H. 2004. Risk benefit analysis of micronutrients. *Food Chemistry and Toxicology* **42**: 1903-1922.

Richardson, D.P. 1997. The addition of nutrients to foods. *Proceedings of the Nutrition Society* **56**: 807-825.

Richardson, D.P. 2007. *Nutrition, healthy ageing and public policy.* International Alliance of Dietary Supplements Associations, Brussels, Belgium. Available at: www.iadsa.org

Sandstrom, B. 2001. A framework for food-based dietary guidelines in the European Union. *Public Health Nutrition* **4**(2A): 293-305.

Scientific Committee on Food and Scientific Panel on Dietetic Products, Nutrition and Allergies. 2006. *Tolerable upper intake levels for vitamins and minerals.* European Food Safety Authority, Parma, Italy. Available at: www.efsa.europa.eu

Shrimpton, D. H. 1997. Vitamins and minerals: a scientific evaluation of the range of safe intakes. Council for Responsible Nutrition, UK. Available at: www.crn-uk.org

Shrimpton, D.H. and Richardson, D.P. 2003. *Vitamins and minerals: an overview of benefits and safety.* Health Food Manufacturers' Association, UK. Available at: www.hfma.co.uk

Smith, A.D. 2007. Folic acid fortification: the good, the bad, and the puzzle of vitamin B-12. *American Journal of Clinical Nutrition* **85**(1): 3-5.

UK Expert Group on Vitamins and Minerals (EVM). 2003. Safe upper levels for vitamins and minerals. Available at: www.food.gov.uk/multimedia/pdfs/vitmin 2003.pdf

UK Office for National Statistics. 1998. *The national diet and nutrition survey (NDNS): people aged 65 and over.* The Stationery Office, London, UK.

UK Office for National Statistics. 2000. *The national diet and nutrition survey (NDNS): young people aged 4 to 18 years.* The Stationery Office, London, UK.

UK Office for National Statistics. 2003. *The national diet and nutrition survey (NDNS): adults aged 19–64 years.* The Stationery Office, London, UK.

Vieth, R. 2006. Critique of the considerations for establishing tolerable upper intake levels for vitamin D: critical need for revision upwards. *Journal of Nutrition* **136**: 1117-1122.

Vieth, R., Bischoff-Ferrari, H., Boucher, B.J. *et al.* 2007. The urgent need to recommend an intake of vitamin D that is effective. *American Journal of Clinical Nutrition* **85**: 649-650.

World Bank. 2004. *Enriching lives. Overcoming vitamin and mineral malnutrition in developing countries*: 1–73. The International Bank for Reconstruction and Development, Washington DC, USA.

World Health Organisation. 2002. Principles and methods for the assessment of risk from essential trace elements. *Environmental Health Criteria* **228**: 1-60 WHO, Geneva, Switzerland.

World Health Organisation. 2005. *A model for establishing upper levels of intake for nutrients and related substances: report of a joint FAO/WHO Technical Workshop on nutrient risk assessment* 1-357. Available at: www.who.int/ipcs/highlights/nutrientproject_may18/en/index.html

Appendix 1. Calculation of the population safety index (PSI) to characterise risk

Nutrient and unit of weight	Tolerable upper intake level (UL)[a]	Mean highest intake from foods only[b]	Potential intake from water (IW)[c]	Reference labelling value (RLV)	Population safety index (PSI)[d] Average PSI in bold
GROUP B					
Niacin (as nicotinamide) mg	M 900	47	0	16	53.3
	F 900	32	0	16	54.3
					53.8
Vitamin E (mg as α-tocopherol equivalent)	M 300	14.5	0	12	23.8
	F 300	12.5	0	12	23.9
					23.9
Vitamin C (mg)	M 2000*	209	0	80	22.4
	F 2000*	187	0	80	22.7
					22.6
Vitamin B_6 (mg)	M 25	5.9	0	1.4	13.6
	F 25	3.8	0	1.4	15.1
					14.4
Vitamin D (μg)	M 50	11.2	0	5	7.8
	F 50	9.6	0	5	8.1
					8.0
Molybdenum (μg)	M 600	210*	20	50	7.4
	F 600	–	–	–	–
Selenium (μg)	M 300	92*	0	55	3.8
	F 300	–	–	55	–
					3.8
Phosphorus (mg)	M 4000*	2710	10	700	1.8
	F 4000*	1806	10	700	3.1
					2.5
GROUP C					
Iron (mg)	M 45*	26.4	0.4	14	1.3
	F 45*	18.9	0.4	14	1.8
					1.6
Iodine (μg)	M 600	407	30	150	1.1
	F 600	–	–	150	–
					1.1
Copper (mg)	M 5	3.3	1	1	0.7
	F 5	2.8	1	1	1.2
					1.0
Calcium (mg)	M 2500	1774	300	1000	0.4
	F 2500	1296	300	1000	0.9
					0.7
Zinc (mg)	M 25	21.3	1	10	0.3
	F 25	12.8	1	10	2.1
					1.2
Vitamin A (preformed retinol μg)	M 3000	2782	0	800	0.3
	F 3000	2521	0	800	0.6
					0.5
Magnesium (mg)[e]	M/F 250	See note	See note	375	**0.7**
Folic acid (μg)[f]	M 1000	203	0	400	2.0
	F 1000	153	0	400	2.1
					2.1

M, male; F, female.
[a] UL as established by SCF/EFSA where available, otherwise the IOM (indicated by *).
[b] Data drawn from Table 3; 97.5 percentile, food only.
[c] Data drawn from SCF/EFSA and EVM opinions and used in ERNA (2004).
[d] The lowest PSI for adult men is used in the risk characterisations shown in Table 5.
[e] UL for dissociable magnesium salts in supplements only.
[f] The UL established by the SCF for folic acid is 1000 μg/day for supplementation.

Mining whole grains for functional components

Julie M. Jones

Department of Family, Consumer and Nutritional Sciences, College of St. Catherine, 2004 Randolph Ave., St. Paul, MN 55105, USA. Tel. 651 690 6664. Fax 651 690 6351. E-mail jmjones@stkate.edu

Abstract

Whole grains contain a myriad of functional components that work both alone and in combination to benefit overall health and reduce the risk of disease. They contain long-recognised health-promoting components such as dietary fibres. They also contain traditional micronutrients such as the tocopherols and tocotrienols and magnesium, which offer well-documented health benefits. However, whole grains contain many compounds that have only recently been recognised as important nutritionally. In some cases the types and roles of fibre from whole grains have been explored more fully. For example, the viscous fibres such as β-glucan from barley and oats or arabinoxylans from wheat and rye are associated with blood glucose attenuation and cholesterol lowering. In addition, whole grains contain some newly-enfranchised fibre constituents, including resistant starch and prebiotic fibres such as inulin. Adding to this is the recognition that many of the phytochemicals in whole grains, including the sterols and stanols, lignans, numerous polyphenolics, sphingolipids, carotenoids and betaine and other methyl donors, work together to promote health. Grain-specific compounds, such as the powerful antioxidant oryzanol from brown rice and avenanthramides from oats, each make important dietary contributions. The identification of these compounds and the understanding of how they work, both singly and in tandem, helps to build the case for inclusion of whole grains in the diet. The message regarding their benefits needs to be disseminated, as large parts of many Western populations fail to even come close to meeting the recommended three servings of whole grain foods per day. Consumers and health professionals need to understand that the benefits of whole grains are not limited to their fibre content, and that the grain is a unique package where the whole is greater than the sum of its parts.

Keywords: whole grains, phytochemicals, antioxidants, human health, glucan, betaine, lignans

1. Introduction

Whole grains have long been regarded as foods that are useful for health. Whole grain bread was mentioned in the Bible (Ezekiel 4:9–17). The Egyptians, Greeks and Romans used bread and whole grains. Barley's genus (*Hordeum vulgare*) is derived from *'hordearii'* meaning gladiator, because it was the food of the Roman gladiators. The philosopher Plato imagined an ideal world where men would live to a ripe old age and eat whole grains grown locally. In the last century the US fundamentalist preacher Sylvester Graham damned the evils of sin and refined grains and promoted the health benefits of whole grains. To this day, graham flour is another name for whole grain flour.

In the 100 years following the exhortations by Dr. Graham to eat more whole grains, they continued to be recognised as nutritious and a good source of fibre, but were not deemed as being important for a healthy diet. However in the last 15 years, many studies have invigorated interest in the nutritional importance of whole grains. Numerous epidemiological studies on a variety of different cohorts indicate that those who eat the most whole grains have reduced risk of a number of different diseases and have reduced mortality at any age.

The link between whole grains and a reduced risk of developing a variety of diseases has piqued an interest in determining the mechanism of action of whole grains and their components, either working alone or in tandem. These components may be both macro-constituents such as fat or dietary fibre, as represented by the proximate composition of common grains in Table 1, or they may be micro-constituents such as minerals or phytochemicals. They may be traditional nutrients and dietary components that have been long associated with both health and whole grains, such as vitamin E and dietary fibre in its many forms. Alternately, they may also be phytochemicals such as ferrulic acid or lignans, long recognised as whole grain components, but only recently recognised as having an impact on health.

Food Science and Technology Bulletin: Functional Foods **4** (7) 67–86
DOI: 10.1616/1476-2137.15050. Published 30 November 2007
ISSN 1476-2137

Table 1. Proximate composition for some common US grains

	Composition (g/100g)			
	Red wheat	White wheat	Barley	Oats
Protein	10.4	11.31	12.5	16.9
Carbohydrate	72.4	75.9	73.5	66.3
viscous	1.5	1.2	3.5	4.2
non viscous	11.0	11.3	13.8	6.4
Total fat	1.6	1.7	2.3	6.9
monounsaturated	0.2	0.2	0.3	2.2
polyunsaturated	0.7	0.7	1.1	2.5

This paper is about whole grain cereals and pseudocereals. It is therefore important to delineate specifically the grains under discussion. Cereals are the seed heads of grasses from the Poaceae (also known as Gramineous) family. Pseudocereals are seed heads of a number of different species of plants but they do not belong to the grass family. A list of cereals and pseudocereals was compiled by the American Association of Cereal Chemists International (AACC International) Task Force on Whole Grains (AACC International 2007; Table 2). The Task Force's rationale for including pseudocereals stated that the grain heads of pseudocereals were used in the same traditional ways that cereals were used in the diet, such as in the making of bread, starch staples and side dishes. Furthermore, they argued that the overall macronutrient composition (proportions of carbohydrate, protein and fat) of both cereals and pseudocereals is similar. They also noted that while legumes and oilseeds have important nutritional contributions, their macronutrient content is quite different from cereal grains. (Note: the list in Table 2 does not include legumes or oilseeds.)

Clarity in the definition of whole grains is critical as researchers seek to understand their physiological role and their components. A committee of experts from AACC International put forward the following technical definition of whole grains: "Whole grains shall consist of the intact, ground, cracked or flaked caryopsis, whose principal anatomical components – the starchy endosperm, germ and bran – are present in the same relative proportions as they exist in the intact caryopsis." (AACC International 1999). A consumer-friendly definition of whole grains was developed in conjunction with members of the AACC International committee by the Whole Grains Council, a non-profit organisation that promotes whole grain usage to consumers and encourages the production of whole grain products by industry. It reads as follows: "Whole grains or foods made from them contain all the essential parts and naturally-occurring nutrients of the entire grain seed. If the grain has been processed (e.g. cracked, crushed, rolled, extruded and/or cooked), the food product should

Table 2. Cereals and pseudocereals (AACC International 2007)

Cereals	Scientific name
True cereals	
Wheat including spelt, emmer, farro, einkorn, kamut, durums	*Triticum spp.*
Rice, African rice	*Oryza spp.*
Barley	*Hordeum spp.*
Corn (maize, popcorn)	*Zea mays*
Rye	*Secale cereale spp.*
Oats	*Avena spp.*
Millets	*Brachiaria spp.; Pennisetum spp.; Panicum spp.; Setaria spp.; Paspalum spp.; Eleusine spp.; Echinochloa spp.*
Sorghum	*Sorghum spp.*
Teff (tef)	*Eragrostis spp.*
Triticale	*Triticale*
Canary seed	*Phalaris arundinacea*
Job's tears	*Coix lachrymal-job*
Fonio, black fonio, Asian millet	*Digitaria spp.*
Wild rice	*Zizania aquatica*
Pseudocereals	
Amaranth	*Amaranthus caudatus*
Buckwheat, tartar buckwheat	*Fagopyrum spp.*
Quinoa	*Chenopodium quinoa Willd* – is generally considered to be a single species within the Chenopodiaceae

deliver approximately the same rich balance of nutrients that are found in the original grain seed." (Whole Grains Council 2003).

The rather substantial body of evidence gathered in the last 15 years showing that the ingestion of whole grains has a positive impact on health, has led government bodies and health promotion organisations around the world to recommend an increased whole grain intake. For example, the 2005 US Dietary Advisory Committee (USDA 2005) has recommended that consumers make 'half their grains whole.' For adults on a 2000-calorie diet, this suggests that three of the six grain servings should consist of whole grains. If these servings were all whole grain bread, this would translate as a minimum of 48 g of whole grains per day. Calculations trying to derive the actual number of grams of whole grain needed per day were based on epidemiological data, which showed that those who ingested 2.7 servings had lower risks of a variety of diseases than those who ingested 1.0 serving or less per day. In many cases, the whole grain servings were achieved through the use of whole grain bread. The following assumptions were made. A slice (1 oz or 30 g) of 100% whole grain bread would contain 16 g of whole grain per slice. Thus, this and similar calculations have led to the recommendation of at least 48 g of whole grains per day.

Despite numerous recommendations by the USDA and other groups, such as the Department of Health and Human Service's Healthy People 2010 challenge, 90% of North Americans fail to meet the whole grain recommendations (Cleveland *et al.* 2000). The European Union Healthgrain platform (see www.healthgrain.org/pub/) and organisations such as Australia's GoGrains have similar recommendations (see www.gograins.com.au). Thus, there are strategies in many countries to increase whole grain consumption.

2. Functional components of whole grain: macro constituents

2.1 Proteins of whole grains as functional components

The proteins of whole grains are nutritionally functional, despite the limiting concentrations of certain essential amino acids, specifically lysine, methionine and tryptophan. They contain high levels of non-essential amino acids that are responsible for the down-regulation of insulin and up-regulation of glucagons, which in turn stimulate adenyl cyclase, resulting in increased levels of cyclic adenosine-3,5-monophosphate (c-AMP) concentrations. The increase in c-AMP is important because it down-regulates the synthesis of a number of enzymes required for *de novo* lipogenesis and cholesterol synthesis, up-regulates key gluconeogenic enzymes and low-density lipoprotein (LDL) cholesterol receptors, and decreases insulin-like growth factor (IGF-1) activity (Krajcovicova-Kudlackova *et al.* 2005). Furthermore, grain proteins such as wheat gluten with a low ratio of lysine/arginine help to reduce the risk of developing hypercholesterolaemia and atherosclerosis through their effects on nitric oxide production (Wang *et al.* 1994). Thus, the protein composition of whole grains may play a role in reducing atherosclerosis and the risk factors for other chronic diseases.

2.2 Fats and carbohydrates of whole grains as functional components

Whole grain foods are low in fat, usually containing just 3–5%. The fat is predominantly unsaturated, consisting largely of linoleic and oleic acids. These have not been found to contribute to cholesterol elevation. In addition, grains contain some unusual fatty components. These may work in tandem with some of the fat-soluble constituents such as tocopherols and tocotrienols to attenuate blood lipids. These will be discussed later in this review.

Whole grain foods are important dietary contributors of carbohydrates, especially starch and dietary fibres of many types – soluble, viscous and insoluble. Some whole grain soluble fibres, inulin and related compounds, and resistant starch were officially recognised in a recent iteration of the dietary fibre definition (AACC International 2001). Inulin and other fructo-oligosaccharides (fructans) have degrees of polymerisation of between 3 and 20 sugar units and can function as prebiotic fibres. The inulin content of some common foods is listed in Table 3. Most whole grains, depending on their variety and degree of processing, contain resistant starch. This starch, which escapes digestion in the small intestine, is fermented by the organisms in the large bowel and thereby contributes some of the physiological functions of dietary fibre.

Table 3. Inulin in foods

Food	Amount of inulin (mg/100g)
Jerusalem artichoke	17–20
Chicory	15–20
Garlic	16
Onions	1–10
Artichokes	2.5
Raisins	1–2
Wheat, barley, rye	0.6–1
Bananas	0.7

3. Whole grains, dietary fibre and health

A view long-held by many health professionals is that the cereal fibre component of whole grains was the major reason to recommend their consumption. Dietary fibre has been defined by AACC International as "the edible parts of plants or analogous carbohydrates that are resistant to digestion and absorption in the human small intestine with complete or partial fermentation in the large intestine. Dietary fibre includes polysaccharides, oligosaccharides, lignin and associated plant substances. Dietary fibres promote beneficial physiological effects including laxation, and/or blood cholesterol attenuation, and/or blood glucose attenuation."

Whole grain foods do indeed contribute much of the diet's total fibre intake and constitute an important cereal fibre along with brans derived from whole grains. The average dietary fibre level in cereal grains varies from 1.2 to 13% (Table 4). Brown rice has the lowest amounts of fibre of all the common grains and barley the highest, with some barley varieties having very high levels (Kaukovirta-Norja 2006; USDA nutrient database, see www.nal.usda.gov/fnic/cgi-bin/nut_search.pl). In addition to significant variability in the amount of dietary fibre across grain species, varieties and cultivars, there is also variability in the type of fibre contained by the grain. Table 4 gives the dietary fibre levels of various common grains per 100 g, which is used as a label basis in much of the world. It also gives the amounts that would be found in commonly-labelled grain serving sizes according to the

Table 4. Fibre content of various common whole grains (WG) in g/100 g and legal labelled serving amounts in North America (from USDA Nutrient Database, Nutrition Labeling and Education Act, and General Mills Inc.)

	Fibre content (g)			
	Brown WG rice	Corn meal	Whole oats	WG wheat
100 g serving	1.8	7.3	10.3	12.2
55 g serving	0.99	3.65	5.67	6.71
30 g serving	0.54	2.19	3.09	3.66

US Food and Drug Administration (FDA) and Health Canada regulations. These include the 55 g serving, which is allowed for some bread servings, and a 30 g serving, which is commonly used for cereals, breads and most grain-based foods.

Fibre intakes in Western countries are far below recommended levels. The current US Daily Reference Intakes recommend that adult males between 19 and 50 years ingest 38 g of dietary fibre per day and females 25 g per day. The average intake of dietary fibre in the US is about 12–13g per day, making intakes one-half to one-third of the recommendation. The 12–13 g of fibre in the diet is largely derived from refined bread products, French fried potatoes and other low fibre foods.

The dietary fibre levels are low since 90% of Americans fail to eat the recommended levels of either breads and cereals or fruit and vegetables, the dietary fibre-containing foods. In a dietary pattern where half the grains are whole and a person eats a whole grain bran cereal and two slices of whole grain bread, the fibre intake from grains would be about 13–17 g per day, or roughly one-half to one-third of the recommended intake. Thus, if consumers follow current dietary recommendations from MyPyramid (a personal healthy eating plan promoted by the USDA, see www.mypyramid.com) and include at least three servings (90 g) of whole grain and 8–13 servings (approximately 400 g) of fruit and vegetables, dietary fibre levels would reach 38–50 g per day.

Fibre intakes in Europe are somewhat higher than in the USA, with levels in northern Europe being higher than parts of eastern and southern Europe (Jones 2004). Median fibre intake in women in the Harvard Nurses' Study in the highest quintile was 24.7 g per day (Liu *et al.* 2000b). In the European Prospective Investigation into Cancer and Nutrition (EPIC) study, median fibre intakes ranged from 12.64 g per day for women in the lowest quintile to 31.91 g for women in the highest quintile. The ranges for men were 12.77 and 35.61 g per day, respectively (Bingham *et al.* 2003).

Studies indicate that consumers ingesting cereal fibres from bran and whole grains have a reduced risk of a variety of diseases, such as cardiovascular disease and diabetes. The risks for coronary heart and peripheral artery disease were reduced from 25 to 35% for the highest whole grain eaters (Erkkila *et al.* 2005; Jensen *et al.* 2004; Merchant *et al.* 2003; Pereira *et al.* 2004). Diabetes risk was reduced from 21% in the Iowa Women's Health study for those in the highest quintile of whole grain ingestion, to 35% for Finnish men and women ingesting the most whole grain. When the studies are assessed in the aggregate, risk reductions averaged about 25% for those who ingested the most whole grain (Stevens *et al.* 2002; Montonen *et al.* 2003; Meyer *et al.* 2000; Liu *et al.* 2000b; Buyken *et al.* 1998; Ylonen *et al.* 2003). Insulin resistance was reduced by 10% in a small intervention study of overweight adults (Pereira *et al.* 2002; Liese *et al.* 2003; Hung *et al.* 2003); stroke was reduced by 36% (Liu *et al.* 2000a).

Whole grain intake has also been related to a reduced risk of elevated blood pressure and weight gain and a strengthened immune system. Hypertension was reduced by up to 40% in the quintile with the highest fibre and whole grain intake, and β-glucan from oats was found to lower blood pressure in obese subjects (Maki *et al.* 2007; Alonso *et al.* 2006; Lairon *et al.* 2005; Keenan *et al.* 2002). A tendency to gain weight over time and a likelihood of showing elevated measures of body weight, such as increased body mass index (BMI), waist-hip ratio or waist circumference, were also reduced in cohorts eating the most whole grain (Barton *et al.* 2005; Lairon *et al.* 2005; Bazzano *et al.* 2005; Koh-Banerjee *et al.* 2004; Lui *et al.* 2003). In addition, specific whole grain fibres such as β–glucan from barley and oats have been shown to increase the immune response (Ramakers *et al.* 2007; Ferguson *et al.* 2001).

Cereal fibres and whole grains have also been associated in animal and human studies with lower risks of some, but not all, cancers. Numerous plausible mechanisms have been suggested for this. In the upper digestive tract, cancer risk is reduced by as much as 50% for the tertile eating the most whole grain and cereal fibre (Campos *et al.* 2005; La Vecchia *et al.* 2003). The link between whole grain and/or dietary fibre intake and colorectal cancer is far less clear. While animal studies consistently show inverse relationships between fibre intake and colon cancer, human studies fail to show a consistent relationship. Such inconsistency is demonstrated in a recent meta-analysis, when it was demonstrated that the link between dietary fibre and a reduced risk of colon cancer was no longer significant when other dietary and lifestyle factors were considered in the regression (Park *et al.* 2005). Further research is needed to clarify the role of whole grains and cereal fibre in colon cancer and other cancers.

In the following section, the roles of fibre from whole grains will be explored further. The longest recognised

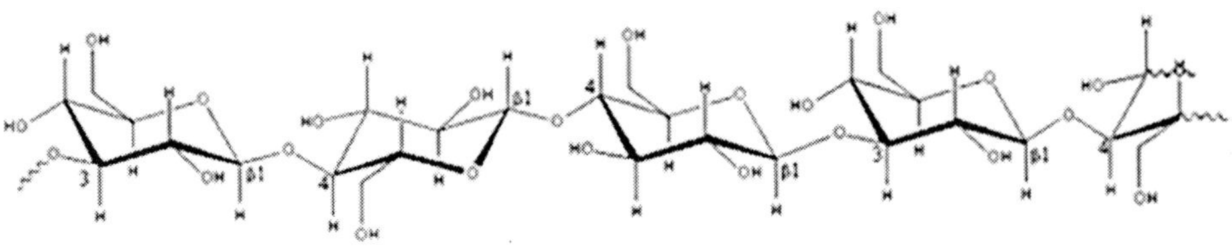

Figure 1. Structure of β-glucan

relationship between whole grain and its fibre component and a health outcome—gut health—will be discussed first. A discussion of the relationship between whole grain and the prevention of other diseases will follow.

3.1 Whole grain fibre and gut health

Adequate dietary fibre, especially that from cereal and whole grains, is associated with increased laxation and improved gut health. There are a number of reasons for this. Cereal fibres, especially insoluble ones, increase both faecal water retention and faecal bulk. More bulky, fluid stools are less abrasive and move through the gut faster, resulting in less constipation and straining at the stool. With less straining, the incidence of haemorrhoids is also reduced. Faster transit times mean that a carcinogen present in the faecal mass will have less time to act on the colonic wall (Chen *et al.* 1998). Increased faecal bulk also means that potential carcinogens are diluted and may be physically separated from colonic cells. Another important way that cereal fibre can improve gut health is through the binding of bile acids. Bound bile acids are far less likely to convert to potentially cancer-promoting secondary bile acids (van Bennekum *et al.* 2005; Grasten *et al.* 2000).

Dietary fibre also affects colonic microflora. These organisms ferment fibrous components to yield short-chain fatty acids. This, in turn, acidifies the content of the gut and lowers colonic pH (Wong *et al.* 2006). The more acidic colonic milieu:

- Reduces the solubility and availability of free bile acids (Kendall *et al.* 2004)
- Neutralises harmful nitrogenous compounds and ammonia
- Favours beneficial enzymatic reactions
- Promotes beneficial microflora. With fewer putrefying bacteria and more beneficial organisms, the immune system is thus strengthened.

The fermentation of dietary fibre results in short-chain fatty acids – butyrate, propionate and acetate. These exert their effects both locally and systemically. Butyrate, the preferred fuel of healthy colonic cells, supports the growth of healthy cells and causes abnormal cells to undergo apoptosis. This reduces the chance for normal colonic cells to mutate to aberrant forms, which could result in polyps and subsequently adenomas. Butyrate not only possesses antineoplastic activity, but it also has been shown to up-regulate tumour suppressor enzymes (Nguyen *et al.* 2006).

All of these mechanisms may contribute to whole grain's antineoplastic activity in the gut (Lim *et al.* 2005). In short, whole grain fibre works in a number of ways that are favourable for gut health.

3.2 Whole grain fibre, blood lipids and glucose response

Cereal fibre from whole grains has also been linked to an improvement in blood lipids and in the glucose response. Dietary fibre from whole grains lowers serum cholesterol by two mechanisms. First, the short-chain fatty acids produced in the gut from fibre fermentation are absorbed and carried to the liver. There they inhibit the same cholesterol-synthesising enzyme found in popularly prescribed statin drugs. Second, viscous fibres such as the β-glucans (Figure 1) from oats and barley form a gel in the gut that binds cholesterol. This in turn inhibits cholesterol absorption by as much as 20% and promotes its excretion and the excretion of bile by up to 144% (Ellegard and Andersson 2007). The excreted bile must be replaced for its use in digestion. In order to synthesise bile, the body uses existing cholesterol. This, in turn, lowers the total pool of body cholesterol.

The rich β-glucan content of oats and barley, compared to other common grains (Table 5), is especially effective for lowering serum cholesterol. Daily oat or oat bran consumption lowers total and LDL-cholesterol an average of 5–8% (Ripsin *et al.* 1992; Karmally *et al.* 2005; Poppitt *et al.* 2007). Numerous studies associating β-glucans and cholesterol lowering enabled health claims to be authorised by the US FDA for both oat and barley β-glucans (FDA, HHS 2006).

Whole grains such as wheat, rye and corn contain no β-glucan. Yet, they are theorised to lower serum cholesterol, since epidemiological data show an association between the ingestion of whole grains and cereal fibre and reduced risk of coronary heart disease (Erkkila *et al.* 2005; Jensen *et al.* 2004; Merchant *et al.* 2003; Pereira *et al.* 2004). In

Table 5. β-glucan content of various grains

Grain	β-glucan content (% dry weight)
Oatmeal	3.5
Oat bran	7.0
Barley	5–6
Rye	1–3
Wheat	<1
Corn	<1

many of these studies, the majority of whole grain intake is from wheat and other grains that have no β-glucan. Components such as the pentosan fractions, primarily arabinoxylan, have been shown through recent studies to also lower cholesterol and to contain factors associated with a reduced risk of coronary disease. For example, one study showed that feeding of arabinoxylan from wheat for six weeks to diabetic subjects lowered the serum lipid risk factors triglyceride and apolipoprotein A-1 (Garcia *et al.* 2006).

Whole grain fibres in the gut also produce beneficial changes that have an impact on glucose and insulin responses in the body. Oat and barley β-glucans have been shown to blunt both of these responses (Behall *et al.* 2006). Data from David Jenkins' lab (Jenkins *et al.* 2002) indicates that for each 50 g carbohydrate portion, the addition of 1 g of β-glucan reduces the glycaemic index by 4 U. In epidemiological and intervention studies a more blunted glucose response is suggested to contribute to reduced insulin resistance and a reduced risk of developing metabolic syndrome and diabetes for those ingesting high cereal fibre and whole grain (Poppitt *et al.* 2007; Weickert *et al.* 2006).

Soluble pentosans found in wheat, rye, rice, millet and corn may also play a role in normalising the blood glucose response (Lopez *et al.* 1999). For example, feeding of arabinoxylans decreased postprandial serum glucose and serum insulin and significantly improved whole-body insulin sensitivity in subjects with impaired glucose tolerance (Garcia *et al.* 2006). Thus, fibres from various whole grains appear to be associated with improved glucose tolerance and reduced risk of insulin resistance and diabetes.

4. Whole grain functional fat-soluble components

Whole grains contain many functionally-active, fat-soluble compounds such as sterols, stanols and sphingolipids. In addition, particular grains may contain some unusual fatty acids and fat-soluble components that are grain specific.

4.1 Whole grain and plant stanols/sterols and alcohols

Whole grains contain many types of sterols and stanols. In wheat and rye, sitosterol is a dominant form (Iwatsuki *et al.* 2003; Hakala *et al.* 2002). In corn, sitostanol is dominant (Moreau *et al.* 2000). Whether in the free or bound form, they are found in highest concentrations in the aluerone layer (Ostlund *et al.* 2002).

Sterols and stanols have been shown to have lipid-lowering properties (Ostlund and Lin 2006). In a European study of over 22 000 the mean total serum cholesterol concentration for men in the highest quintile of plant sterol intake (mean intake 463 mg/day) was 0.25 mmol/L lower and LDL-cholesterol was 0.14 mmol/L lower compared to those in the lowest quintile (mean intake 178 mg/day); the corresponding figures for women were 0.15 and 0.13 mmol/L, respectively (Andersson *et al.* 2004). In this UK sample (n = 24 798), the mean intake of total plant sterols was 300 mg/day for men and 293 mg/day for women. Whole grain bread and other cereals was one of three major food categories providing plant sterols, accounting for just under 20% of the total daily intake (Klingberg *et al.* 2007).

Use of sterol- and stanol-fortified food products and margarines can boost daily consumption to 2–3 g/day. Studies show that the use of these foods in free-living populations decreases LDL-cholesterol concentrations by 9–14% (Kerckhoffs *et al.* 2002). It is more difficult to determine the precise contribution that sterols from whole grains make to the overall health benefits of whole grains. One study documented increases in blood sterol levels with whole grain ingestion, but failed to show an impact on serum lipids during the duration of the study (Clifton *et al.* 2004). It is not known if the concentrations achieved were too low to have an impact, or if the lack of effect was due to the short duration of the study.

4.2 Whole grain waxes

Whole grain plant waxes, mixtures of very long chain (20–36 carbon) aliphatic primary alcohols, are also found in whole grains. Wheat, especially wheat bran, is high in policosanol with tetracosanol (C24), hexacosanol (C26) and octacosanol (C28), with the later predominating. The amount in grains depends both on the type of grain and its variety (Irmak and Dunford 2005).

Reports suggest that 5–20 mg/day of mixed C24–C34 alcohols, including octacosanol and triacontanol, lower LDL-cholesterol by 21–29% and raise high-density lipoprotein (HDL)-cholesterol by 8–15% (Hargrove *et al.* 2004). Another study showed that extracts of policosanol (5 mg/day) ingested for eight weeks reduced serum total cholesterol by 12% and LDL-cholesterol levels by 17%, and raised HDL-cholesterol and improved LDL/HDL ratios in those with elevated blood lipids (Mirkin *et al.* 2001). Physiologically these compounds have been reported to reduce risk factors for cardiovascular disease by altering cholesterol absorption and metabolism, improving blood lipid profiles and reducing the oxidation of human cholesterol components (Ortega *et al.* 2006; Menendez *et al.* 2000).

Data on isolated policosanols is promising. For example, octacosanol isolated from wheat germ is associated with an improvement in various markers of physical fitness (Irmak and Dunford 2005). Policosanols have also been hypothesised as being potentially beneficial in reducing the risk of diseases such as Alzheimer's (McCarty 2006). However, the actual contribution of these compounds delivered as part of whole grain foods to health has yet to be researched.

4.3 Whole grain sphingolipids and squalene

4.3.1 Sphingolipids

Sphingolipids and their metabolites are found in the cell membranes of all higher organisms. They are digested throughout the gastrointestinal tract and metabolised to ceramide and sphingosine. These highly bioactive compounds have effects on various cellular functions including regulation of lipid synthesis and lipoprotein metabolism, cell growth, differentiation and apoptosis (Worgall 2007; Merrill *et al.* 1995).

The cellular regulatory role of these compounds may also help prevent atherosclerosis and certain types of cancers (Nilsson and Duan 2006). In terms of atherosclerosis, sphingolipids exert control over rate-limiting enzymes of fatty acid and cholesterol synthesis as well as fatty-acyl-CoA synthases. Studies in animal models suggest that inhibition of sphingolipid synthesis correlates with decreased atherosclerotic lesions and a decreased susceptibility of lipoproteins to deposit on the vessel wall—a key mechanism in the development of the atherosclerotic lesion (Padron 2006).

Sphingolipids are also associated with control of tumour growth and maintenance of normal epithelia. *In vitro* studies with wheat sphingolipids, like their dairy counterparts, show that they cause apoptosis of colon cancer cell lines, reduce the appearance of aberrant colonic crypts and decrease the number of tumours in mice treated with a carcinogen (Sugawara *et al.* 2002; Merrill *et al.* 1995). In addition, supplementation with rice bran glucosylceramide repaired skin barrier function in mice with chronic skin perturbation (Tsuji *et al.* 2006).

The intake of these compounds in a typical Western diet is estimated to be 0.3–0.4 g/day, but the amount contributed by plant sources may be as little as 50 mg/day (Sugawara and Miyazawa 1999). More research is needed concerning the physiological role of these components from grains, because existing data are mostly from the role of sphingolipids in eggs, dairy and human milk. One rodent study suggests that sphingolipid isomers from plants are absorbed similarly to those from animal products, but also suggests that there are some metabolic differences (Sugawara *et al.* 2003). Nonetheless, dietary sphingolipids from plant sources are thought to influence human health (Berger *et al.* 2003).

4.3.2 Squalene

Squalene is a lipid component of some whole grains and pseudocereals such as amaranth (Martirosyan *et al.* 2007). It is an isoprenoid compound similar in structure to β-carotene. As such, it acts as a quencher of singlet oxygen-inhibiting lipid peroxidation. It may also inhibit some cancers (Smith 2000).

Animal studies show that squalene-containing grains and oils, such as amaranth oil, can modulate cell membrane fluidity and stabilise membranes. It also appears to decrease blood cholesterol by 15–20% in studies with experimental animals (Berger *et al.* 2003). More research is needed on the contribution of grains containing small amounts of this functional component.

5. Whole grain minerals and vitamins

Whole grains are a good source of traditional nutrients including vitamins and minerals. In typical Western diets about 50% of iron, 30% of magnesium and 20% of zinc comes from cereal products (Lewis and Buss 1988). Most of the minerals are in the bran and germ portions, but are tightly bound to the bran, fibre and phytate. These components of grain can impair divalent mineral absorption (Hurrell 2003). However, not all studies show an adverse effect. For example, whole wheat flours in the diets of mice caused no impairment in mineral absorption (Levrat-Verny *et al.* 1999). Nonetheless, special efforts must be made for plant-based diets comprised of predominantly whole grains, especially high phytate ones such as sorghum and millet, to ensure bioavailability of such minerals (Perlas and Gibson 2005).

5.1 Whole grain and iron

Whole grains contain iron, but the amount varies, both by grain type and cultivar. For example, there was a four-fold variation in the iron level in a study of 15 cultivars. The amount of iron varied from 80 to 368 μg/g dry weight (Welch *et al.* 2005).

Iron bioavailability varies as much or more than the amount of iron contained within the grain. Factors in the grain and in the diet affect the bioavailability of non-heam iron (the less bioavailable form) in whole grains (Bach Kristensen *et al.* 2005). Phytate, tannin, phenolics and fibre are inherent components that can inhibit iron absorption (Clydesdale *et al.* 1991). Phytates (inositol hexaphosphates), in particular, have long been associated with poor bioavailability of iron and other divalent minerals. However, not all studies show that phytate from whole grain affects iron bioavailability. In a study with 15 varieties of unpolished rice, phytic acid was not correlated with iron bioavailability (Glahn *et al.* 2002).

The inhibitory effect of phytate is especially problematic for whole grains such as millet and sorghum that also contain tannin, and for consumers whose diets are marginal (Lestienne *et al.* 2005). A variety of human studies have also shown that iron status is not affected by phytate and whole grain intake, provided the diet has a variety of iron sources (Krittaphol *et al.* 2006). Similarly, the mineral status of consumers of plant-based diets in wealthier countries appears adequate where the overall nutrient intake is good (Hunt 2002; see also discussion on phytate, Section 6.3). In contrast, phytate can have a negative impact on iron, zinc and other divalent metals in countries where the overall nutrient availability for at-risk groups is inadequate, or there are secondary factors such as infection that impair iron status (Hambraeus 1999).

Processing can improve iron bioavailability by changing the phytate complex. For example, fermentation allows the phytase to act and increases divalent metal availability. Sourdough processes actually enhance iron and zinc absorption (Lopez *et al.* 2003). Lactic acid production or addition to gruels or other foods improved iron absorption despite the phytate content (Bering *et al.* 2006; Proulx and Reddy 2007). Nixtimalisation of corn for the making of tortillas decreases the phytic acid by nearly 20% (Bressani *et al.* 2004). Most types of heating improved iron bioavailability (Hemalatha *et al.* 2007).

Iron added to fortified whole grain products must be shown to be in a form that is absorbed. Furthermore, food processing and storage conditions should be optimised to maintain the iron as available and, as such, to contribute to iron status.

5.2 Whole grain and magnesium

Whole grains may have over five times more magnesium than refined wheat flour. One serving of whole grain delivers 25–35% of the RDA for magnesium. The magnesium contribution of whole grains may be one factor contributing to the inverse associations between whole grain intake and reduced risk of metabolic syndrome, hypertension, diabetes and coronary heart disease (van Dam *et al.* 2006; He *et al.* 2006; Murakami *et al.* 2005; McCarty 2005; Song *et al.* 2006).

5.3 Whole grain and zinc

Whole grains contain four times more zinc than refined wheat flour. Like iron, the bioavailability of zinc from whole grains can be reduced by phytate and other grain components, and processing affects availability (Sandstrom *et al.* 1987). Pressure cooking was found to decrease zinc bioaccessibility by as much as 60% in rice and millet.

In a study where two meat diets were formulated to have a phytate content equivalent to that of a vegetarian diet, there was lower zinc absorption from the vegetarian diets. However, there was no difference in the fractional absorption, meaning that zinc was unrelated to the phytate (Kristensen *et al.* 2006). In some cases, the addition of gut fermentable fibres actually countered the effect of phytate and increased zinc absorption (Hemalatha *et al.* 2007).

Zinc is important for many reactions in the body and whole grains are an important contributor for those with little or no meat in the diet. Since zinc is involved in more than 100 enzymatic reactions, adequate levels are important for health and disease prevention. There is a strong association between adequate levels of zinc and reduction of diabetes risk. There are numerous mechanisms whereby zinc can modulate insulin activity and affect antioxidant capacity. It is well known that zinc status is decreased in most patients with type 2 diabetes (Yonekura and Suzuki 2005).

5.4 Whole grain selenium

Whole grains contain selenium, if the soil on which they are grown is selenium-rich. Grains can be a very important source of this mineral, especially in countries such as the USA where some areas are selenium-rich and others are not. The blending of grains and the widespread shipping of grain-based foods to many areas helps to distribute this essential micronutrient with its important antioxidant potential to eaters, whether they live in a selenium-rich area or not.

Unfortunately selenium is deficient in at least a billion people worldwide. Many parts of the world have a low soil selenium content. In countries such as Australia and China with dangerously low soil selenium, biofortification through inclusion of selenium in fertiliser products ensures its presence in the grain (Lyons *et al.* 2005; Yang *et al.* 2007). In the UK, the switch from selenium-rich North American wheat to locally-grown wheat decreased dietary selenium intake from over 60 μg/day in the 1970s to 35 μg/day in the 1990s. Since the reference nutrient intakes for adult females and males in the UK are 60 and 75 μg selenium/day, respectively, countries like the UK are also using biofortification (Broadley *et al.* 2006).

5.5 Whole grains and vitamins

Whole grains have long been regarded as a rich source of water-soluble B vitamins such as thiamine, niacin, vitamin B6, folate and vitamin E. The discovery of thiamine as an essential vitamin occurred when birds fed white (polished)

Table 6. Typical tocopherol and tocotrienol content of grains

Grain type	Average amount(s) (mg/kg dry weight)	Predominate type	Notes
Soft wheat and barley	75	Tocols	Highest total tocopherol/ tocotrienol of the grains (Panfili *et al.* 2003)
Hulled/dehulled wheats	33–43	β-tocotrienol	
Maize	45	γ-tocopherol	
Oats	56	α-tocotrienol	
Barley	40	α-tocotrienol	
Waxy barley	60.3–67.6	Tocols	Eastern European varieties
wanubet, wabet, washonubet	18.0–20.1	Vitamin E equivalent	(Ehrenbergerova *et al.* 2006)

rice developed avian polyneuritis. Addition of the rice polishings (the bran and the germ portion) reversed the neurological disorder in the birds. In other words feeding the whole grain provided the disease-preventing components. The B vitamins have long been associated with grains and whole grains and several have been part of national enrichment and fortification programmes. This review will focus on whole grain's contribution of vitamin E isomers.

5.6 *Whole grains and vitamin E*

Whole grains contribute fat-soluble vitamins, particularly the vitamin E family of compounds—tocopherol and tocotrienols. There are four tocopherol isomers (α, β, γ, δ) and four tocotrienol isomers (α, β, γ, δ). α-Tocopherol is the form with the highest vitamin E activity; the other tocopherols have reduced levels of vitamin E. The tocotrienols have little vitamin E activity, but they are excellent antioxidants. Table 6 lists some grains and the predominant types of vitamin E that they contain. Wheat germ is an excellent source, with wheat germ oil having the highest tocopherol content of all the vegetable oils (215.4 mg/100 g). Wheat germ has 11 mg/100g and whole wheat flour has 1.2 mg α-tocopherol equivalents/100 g.

5.6.1 Whole grains, vitamin E and health

Vitamin E compounds are part of the powerful antioxidant capacity of whole grain foods. They block free radicals that have the capacity to attack cell membranes, nucleic acids, proteins and fats. Thus these compounds protect structural and enzymatic proteins, and prevent mutation formation that could occur when DNA and RNA are attacked by free radicals. They also preserve cell structure by keeping cell membranes fluid and functional by protecting membrane and other cellular lipids.

Tocopherols have documented benefits with respect to various diseases. In terms of coronary heart disease, tocopherols from rice bran were shown to lower cholesterol. Rice bran fractions have been shown to reduce cholesterol in a dose-dependent fashion. Reductions of 12–20% were seen when 25–100 mg of tocopherol-rich rice bran was ingested (Qureshi *et al.* 2002). One mechanism for cholesterol reduction is reduced cholesterol synthesis.

In addition to cholesterol lowering, there are other functional aspects that reduce coronary and other disease risk. For example, whole grains have been shown to reduce the oxidation of cholesterol. This is important because oxidised cholesterol particles are more atherogenic than others (Xu *et al.* 2001). Tocopherols from oat and wheat have been shown to alter factors that affect platelet aggregation and modulate vasorelaxation (Katz *et al.* 2003). The vitamin E compounds in foods like whole rice and rice bran have been found to help strengthen the immune system by modulating cell signalling and apoptosis, and to inhibit tumour cell proliferation (Qureshi *et al.* 2000).

Vitamin E intake was significantly associated with a reduced risk of type 2 diabetes (Kanaya *et al.* 2004). Intakes of α-, γ- and δ-tocopherol, and β-tocotrienol were inversely related to the risk of developing this disease (Montanen *et al.* 2004). The highest intake of dietary γ-tocopherol, the form of vitamin E that predominates in food sources, was associated with a 32% reduction in the risk of advanced prostate cancer in a US cohort of nearly 300 000 men aged 51–70 (Wright *et al.* 2007) and a 60% reduction in the risk of breast cancer, compared to those with the lowest consumption.

5.6.2 Whole grains, vitamin E and intake

Many populations do not receive an adequate intake of vitamin E. The 2005 US Dietary Guidelines Advisory Committee listed vitamin E as a nutrient of concern because only 8.0% of men and 2.4% of women met the new Estimated Average Requirement (EAR) of 12 mg from foods alone, in the 1994–1996 Continuing Survey of Food Intakes by Individuals (CSFII). The US Recommended Daily Allowance (RDA) for vitamin E is 15 mg of tocopherol equivalents for both men and women (Gao *et al.* 2006). Most Western populations get most of their vitamin E from vegetable oils and cereals (Nielsen and Hansen 2007). Whole grain cereals have the potential to provide a greater proportion of the daily vitamin E intake, especially as some groups are working to decrease fat and oil intake and others are trying to incorporate more whole grains.

5.6.3 Whole grains, vitamin E, forms and processing

Tocopherol levels in the various grains and varieties differ and are affected by processing and growing conditions. In an Italian study, the highest tocol levels were found in soft wheat and barley (approximately 75 mg/kg dry weight). β-Tocotrienol is the main vitamin found in hulled and dehulled wheats (33–43 mg/kg dry weight), and γ-tocopherol predominates in corn (45 mg/kg dry weight). In Illinois-raised corn, γ-tocopherol was also the major form, with levels ranging from 2.4 to 63.3 μg/g dry weight (Kurilich and Juvik 1999). α-Tocotrienol predominates in oat and barley (56 and 40 mg/kg dry weight, respectively; Panfili *et al.* 2003). Table 6 gives some of the measured values.

Heat treatments of whole grains can reduce some of the important tocopherols and other antioxidant components. Heating corn to high temperatures simultaneously decreased levels of γ-tocopherol and γ- and δ-tocotrienol. In some cases over 50% of the original compound was decreased, and there was an accompanying increase in the production of oxidation products (Moreau and Hicks 2006). Steaming and flaking of oat groats resulted in moderate losses of tocotrienols, but not of tocopherols. Autoclaving of grains in the hulls caused increased levels of all tocopherols and all but one tocotrienol (Bryngelsson *et al.* 2002). The tocopherols were unaffected by steaming, but the drum drying of steamed rolled oats resulted in an almost complete loss of both tocopherols and tocotrienols. In breadmaking, tocopherols and tocotrienols are reduced during sourdough fermentation (Liukkonin *et al.* 2003). In the milling and extruding of buckwheat, there is a decrease of about 10% in overall antioxidant capacity, primarily due to loss of tocopherols (Zielinski *et al.* 2006).

6. Whole grain phytochemicals

Whole grains contain an array of phytochemicals with beneficial nutritional properties. These may be antioxidant compounds that work in tandem with vitamin E and other antioxidant vitamins to help to protect against oxidative stress and, therefore, to decrease the risk of many diseases. In addition to their antioxidant properties, whole grain phytochemicals exert their health benefits through multifactorial physiological mechanisms including:

- Enhancement of the immune system
- Facilitation of substance transit through the digestive tract
- Modulation of metabolism and hormones
- Promotion of antibacterial and antiviral activities.

6.1 Whole grain phenolics and polyphenolics

Phenolics and polyphenolics are found in many foods in the plant kingdom, including whole grains. These, along with other grain components, may be involved in reducing diabetes and other chronic diseases (Montonen *et al.* 2004). Over 8000 naturally-occurring compounds are known. They may be classed as simple phenols, which have a single phenolic ring; these include phenolic acids and alkylresorcinols. Polyphenolics are those with two phenolic rings; acid dehydrodimers, lignans, flavonoids and compounds containing three or more subunits are termed tannins. In grains, the phenolic compounds are free or bound as glycosides, esters or in complexes with cell constituents such as polysaccharides or proteins.

Whole grain phenolics are located in the outer layers of the kernel. Predominant cereal polyphenolics include hydroxycinnamic, ferulic, vanillic, and p-coumaric acids. Other common grain phenolics include gallic, benzoic, p-hydroxybenzoic, sinapic and caffeic acids, together with quercetin, catechin, rutin, and quercetin and diol esters. Cereal products, especially wheat bran, provide much of the ferulic acid in the diet (Manach *et al.* 2004). Ferulic acid is among the most potent antioxidant of all grain phenolics. If a significant amount of whole grain and bran is included in the diet, the intake of ferulic acid is often over 100 mg, in comparison to an average diet in which the intake of phenolic is about 25 mg/day (Baublis *et al.* 2000; Clifford 2004).

Highly-coloured whole grains such as blue corn, black rice or purple wheat offer anthocyanin components such as cyanidin 3-glucoside (Abdel-Aal and Hucl 2003). The total anthocyanin contents vary significantly and exhibit a range of 7–3276 μg/g. Some grains, such as red rice and black rice, contain a limited number of pigments, whereas others, such as blue, pink, purple and red corns, have complex anthocyanin profiles (Abdel-Aal *et al.* 2006).

There also are some grain-specific phenolics, such as avenanthramides from oats or oryzanol in brown rice. All of these constituents contribute to the antioxidant, anti-aging, immune boosting, antimutagenic and antiatherogenic potential properties of grains (Alvarez *et al.* 2006a).

Some of the compounds are bound to sterols or alcohols, such as the steryl ferulates, campestanyl ferulate, sitostanyl ferulate, campesteryl ferulate and sitosteryl ferulate. Like the other ferulates, the largest amounts reside in the outer layers. There is almost none within the highly refined endosperm (Hakala *et al.* 2002). Corn, rye and wheat grains contain 6.0–6.5 mg/100 g steryl ferulates.

The oryzanols of brown rice are mixtures of ferulic acid esters of sterol and triterpene alcohols (Luo *et al.* 2005; Akihisa *et al.* 2000). Some studies show them to be much more powerful than vitamin E or ferulic acid by itself. In animal studies, these compounds have been found to reduce cholesterol levels by over 50%. Further, they inhibit cholesterol oxidation and platelet aggregation (Wilson *et al.* 2007; Scavariello and Arellano 1998). Such changes would lower the risk of coronary heart disease. The effects of oryzanol appear to go beyond heart health as it has been shown to modulate pituitary secretion and inhibit gastric acid secretion (Jariwalla 2001; Cicero and Gaddi 2001).

More research is needed to determine the amounts of phytochemicals present in different species, varieties and cultivars, and the effect of various processing techniques. This information should be used to optimise systems to deliver the largest quantity of bioavailable phytochemicals possible.

6.1.1 Whole grain phenolics and processing

Processing can both increase and decrease the level and bioactivity of many grain phenolic compounds. For example, in the making of Finnish rye bread, levels of folate and easily-extractable phenolic compounds increase during germination and sourdough baking (Liukkonen *et al.* 2003; Hietaniemi and Poutanen 2003). In the milling and extruding of buckwheat, there is about a 10% decrease in overall antioxidant, but this was not due to the loss of phenolics (Zielinski *et al.* 2006). Steaming of oats decreased caffeic acid and one of three isomers of avenanthramide, but ferulic acid and vanillin increased. Drum drying caused a large decrease in total cinnamic acids and avenanthramides (Bryngelsson *et al.* 2002).

Consumers and health professionals are aware of the important antioxidant contribution of fruit and vegetables, but are often unaware that whole grains are an important source. Miller *et al.* (2000) compared the *in vitro* antioxidant activity of whole grain cereals, fruits and vegetables. Whole grain breakfast cereals were found to have an antioxidant capacity of between 2200 and 3500 Trolox equivalents per 100 g sample, compared with most common fruits (excluding berries) that have 600–1700 Trolox equivalents/100 g (Table 7). Since consumers fail to ingest either the recommended intake of fruit and vegetables or whole grains, the average daily dietary antioxidant intake falls well below what it should be. Consumers who switch to whole grains will help increase the consumption of these antioxidants in the diet.

Using data published by Miller *et al.* (2000), the potential average dietary antioxidant levels were calculated with the assumption that adults were heeding current recommendations. In that case, they would be getting four 4 oz servings (approximately 400 g) of fruit per day with an average antioxidant level of 5760 Trolox equivalents per day; four 4 oz servings (approximately 400 g) of vegetables per day with an average antioxidant level of 2060 Trolox equivalents per day; and three 1 oz servings (approximately 100 g) of whole grain per day with an average antioxidant level of 3450 Trolox equivalents per day (Miller *et al.* 2000). Thus, there would be a total in take of over 11 000 Trolox equivalents per day, and whole grains would deliver 30–33% of the requirement. Thus, in most diets a switch to whole grains would deliver a significant proportion of the antioxidants required.

Table 7. Comparison of the antioxidant activity of whole grain cereals and fruit and vegetables, per serving and per 100 g

	Antioxidant activity per serving	
	Serving size (g)	Trolox equivalents
Whole grain cereal	41*	1150*
Vegetables	120	540
Fruits	120	1440
Whole wheat bread	50	1000

*Average values for eight whole grain breakfast cereals.

6.2 Whole grains and lignans

Many whole grains contain lignans, which are phenolic lipids in the alkylresorcinol family. They are found in wheat, rye, triticale and barley (Table 8), but not in rice, oats, maize, sorghum or millet. Within the lignan-containing grains, there is a significant difference in the amount both within and among species. For example, one durum wheat variety has only 54 µg/g compared with another durum variety that has 589–751 µg/g (Ross *et al.* 2003). Table 9 shows the dietary lignan content of diets for animals formulated to have the same fibre content, but one is made from wheat and the other from rye. Both the total lignan content and the concentration of the various lignan species are higher in the rye-based than in the wheat-based diet. As a category, whole grain cereals and bran, and oilseeds, especially flax, and legumes vary in terms of the amount of mammalian lignan produced in the colon (Table 10; Thompson *et al.* 1991). However, since many Western diets fail to include legumes and oilseeds in adequate amounts, bran and whole grains can be important contributors. In a Dutch study of menopausal women,

Table 8. Range of alkyresorcinols in various grains

Grain	Range (µg/g)
Wheat	489–1429
Rye	720–761
Triticale	439–647
Barley	42–51

Table 9. Wheat and rye diets formulated with similar fibre content but different types of fibres and different lignans (data from Bach Knudsen *et al.* 2003)

	Units/kg dry matter	
	Diet 1	Diet 2
Bread type:		
wheat crisp	602 g	—
rye crisp	—	593 g
wheat soft	370 g	—
rye soft	—	379 g
Ash	33 g	56 g
Protein (N × 6.25)	119 g	127 g
Fat	68 g	73 g
Total carbohydrates	773 g	699 g
Sugars	24 g	40 g
Starch	529 g	455 g
Dietary fibre (NSP + lignan)	230 g	235 g
Fructans	4 g	18 g
Cellulose	131 g	20 g
β-glucan	12 g	19 g
Arabinoxylans	53 g	127 g
Klason lignin	10 g	32 g
Total lignans	3131 μg	35177 μg
Mammalian lignan precursors	2465 μg	30480 μg
I-LAR	666 μg	4697 μg
secoisolariciresinol	161 μg	754 μg
matairesinol	33 μg	813 μg
lariciresinol	52 μg	1689 μg
pinoresinol	390 μg	2840 μg
syringaresinol	1829 μg	24384 μg

breads and cereals contributed about 7% of the total isoflavone and lignan intake (Kreijkamp-Kaspers *et al.* 2005). The median total lignan intake was 977 μg/day.

Scandinavians boost their lignan consumption because of their significant intake of rye and barley. For example, 100 g of rye crispbread yields 41.8 and 33.0 μg/100 g dry weight secoisolariciresinol and matairesinol, respectively (Pool-Zobel *et al.* 2000). Matairesinol is derived almost exclusively from whole grains. While secoisolariciresinols are derived from whole grains, they are also provided by fruits, especially berries.

6.2.1 Whole grain lignans and cardiovascular disease

The intake of whole grain lignans is associated, in a number of different populations, with lower coronary disease and cardiovascular mortality. In the Zutphen Elderly Study, matairesinol intake was inversely associated with coronary heart disease and cardiovascular disease (Milder *et al.* 2006). Dutch postmenopausal women with the highest lignan intake had lower aortic stiffness (van der Schouw *et al.* 2002). The tertile eating the most lignan had 64% lower lipoprotein a [Lp(a)] levels than the tertile eating the least lignan (Kreijkamp-Kaspers *et al.* 2005). In Nordic populations, the ingestion of rye was associated

Table 10. Mammalian lignan produced from precursors in various food classes (after Thompson *et al.* 1991)

Food category	Mammalian lignan produced per 100 g
Oilseeds	20 461 ± 12 685
Dried seaweeds	900 ± 247
Legumes	562 ± 211
Cereal bran	486 ± 90
Whole grain cereals	359 ± 81
Vegetables	144 ± 23
Fruits	84 ± 22

with lower cholesterol levels. In one study the cholesterol absorption was 9% lower when the starches in the diet were from whole grain rye than when they came from potato and refined-grain bread (Hallikainen *et al.* 2006). The lignan component of rye is suggested as one of the reasons for the lowered risk.

There are a number of ways that whole grain lignans may reduce cardiovascular risk factors. First, whole grain lignans may reduce bile acids. For example, diets high in rye are associated with increased excretion of bile acids by about 125 mg/day over that seen with a low fibre diet (Lundin *et al.* 2004). Second, dietary lignans are altered by colonic bacteria to produce the mammalian lignans, enterolactone and enterodiol. In a study of Finnish men, those with the highest rye and lignan baseline intakes had higher 5.3 nmol/L higher serum enterolactone levels than those with lower intakes of rye. At the completion of the study, those with the highest lignan intake had higher serum enterolactone levels (>30.1 nmol/L) and a 58.8% lower risk of acute coronary events (Vanharanta *et al.* 1999). Enterolactone has been shown to up-regulate liver LDL receptor activity and to interfere with micellar solubilisation of cholesterol in the intestine (Owen *et al.* 2004). The reduction of cholesterol solubilisation decreases the efficiency of cholesterol absorption. Third, epidemiological studies indicate that whole grain lignans may reduce aortic stiffness and decrease Lp(a) and apoprotein-B (apo-B) lipoproteins. All of these changes would have positive effects on coronary disease risk factors.

6.2.2 Whole grain lignans and cancer

An increased risk of a number of cancers appears to be related to the low intake of phytoestrogens, lignans or dietary fibre-lignan complexes typical in many Western diets. Dietary lignan intake in a case-control study carried out in Ontario was associated with a 27% reduction in colorectal cancer risk (Cotterchio *et al.* 2006). In the Zutphen Elderly Study, those with the highest matairesinol intake had a 28% lower risk of cancer and a 19% decreased risk in all-cause mortality (Milder *et al.* 2006). Case-control studies of breast cancer in German, pre-menopausal women showed a protective effect of matairesinol, but not of secoisolariciresinol or total lignans (Linseisen *et al.* 2004). In a US cohort, total lignans were inversely associated with breast cancer in pre-menopausal but not post-menopausal women (McCann *et al.* 2004, 2006). The quartile of pre-menopausal women with the highest intake had a 36% decrease in risk. The reduced risk was only true for oestrogen receptor-negative types of breast cancer. Those in the group with the highest compared to the lowest quartile of lignan intake had lower tumour risk. Higher phytoestrogen intake (both isoflavones and lignans) during adolescence was also associated with a reduced breast cancer risk (Thanos *et al.* 2006). In other types of cancers, there seemed to be no documented effect of lignan intake. There was no relationship between lignan intake and either endometrial cancer in New York women (Zeleniuch-Jacquotte *et al.* 2006) or prostate cancer in Swedish men (Stattin *et al.* 2004).

For those cancers that appear to be reduced by whole grain and lignan intake, there are several postulated mechanisms regarding effects on hormone metabolism and cancer cell growth (Adlercreutz 1990). The lignin-lignan dietary fibre complex of foods such as rye seems to influence sex hormone and bile acid metabolism by partial interruption of the enterohepatic circulation and alteration of intestinal metabolism, and by increasing faecal excretion of these compounds (Lundin *et al.* 2004). In one study the ingestion of high fibre, whole grain rye bread resulted in nearly 400 mg more bile acid being excreted. Furthermore, urinary enterolactone excetion was 5- and 10-fold higher in men and women eating whole grain rye bread than in those eating wheat bread, respectively (Juntunen 2000).

6.3 *Whole grain phytates*

Whole grains contain phytate. As has been seen with other grain components, the amount varies both with the grain type, the variety and the growing conditions. In wheat alone, concentrations of myo-inositolhexaphosphate (phytate) varied from 8.6 to 26.1 μmol/g dry weight.

The potential adverse effects of phytate on mineral absorption have been addressed in Section 5. Phytate also can adversely affect availability of carbohydrate from whole grain components. In experimental animals, it decreased the rate of digestion of raw wheat starch by 50% (Yoon *et al.* 1983). For those with inadequate calories or otherwise marginal diets this effect of phytate may not be beneficial. However, in this age of caloric surfeit and rampant diabetes, carbohydrate that is less available is advantageous.

Other potential health benefits of phytate have been elucidated. Whole grain brown rice versus milled rice lowered the glycaemic response by as much as 35% in diabetic subjects. Phytic acid was ascribed as one of many possible reasons for this effect (Panlasigui and Thompson 2006). Rice bran-derived phytate, along with a high cholesterol diet, prevented the expected rise in cholesterol (Jariwalla 1999). Through chelation of iron and copper, phytate might reduce oxidation of cholesterol. This also could make iron less available (Fleming *et al.* 2002) in the large bowel and might promote the health of other cell components and, in turn, reduce risk. Both the bran and phytate of rice and wheat have been shown in animal models to increase natural killer cell activity in the colon, to promote cell apoptosis and differentiation, and to favourably affect colon morphology. All these could have an anti-neoplastic role but there have been no human studies to verify these effects (Zhang *et al.* 2005; Fox and Eberl 2002; Jariwalla 2001; Jenab and Thompson 2000; Nielsen *et al.* 1987). Phytate intake in the Nurses Health Study II was associated with a reduced risk of kidney stone formation (Curhan *et al.* 2004).

With phytate as with many of the other functional phytochemicals, more research is needed in order to define the role of phytate as part of a whole grain food and as a component of various types of diets.

6.4 *Whole grain methyl donors*

Whole grains contribute choline (phosphatidylcholine) and betaine (trimethylglycine; Table 11). The phospholipid choline is important for normal membrane function, acetylcholine synthesis and methyl group metabolism. Intakes may be below those considered optimal for health (Fischer *et al.* 2007). Betaine is an osmolyte and methyl donor. As an osmolyte, betaine protects cells, proteins and enzymes from environmental stress (e.g. low water, high salinity or extreme temperature). As a methyl donor, betaine participates in the methionine cycle and in nucleic acid metabolism. Inadequate dietary intake of methyl groups was found to result in elevated plasma homocysteine concentrations, decreased S-adenosylmethionine concentrations and altered fat metabolism that could potentially lead to dislipidaemia (Craig 2004). Betaine in some studies lowered homocysteine (Alfthan *et al.* 2004). There was concern in another study that it also raised LDL levels (Olthof *et al.* 2005).

Table 11. Whole grains, choline and betaine (data from Zeisel *et al.* 2003)

Source	Choline mg/100 g	Betaine mg/100 g
Wheat germ	152	1241
Wheat bran	—	1339
Whole wheat bread	—	201

6.5 Whole grain carotenoids

The carotenoids are minor constituents of grains found in the outer layers of the grain. In fact, the bran and germ layers of wheat had 4-fold more lutein, 12-fold more zeaxanthin and 2-fold more β-cryptoxanthin than the endosperm fractions (Adom *et al.* 2005).

Corn is one of the few grains that contain vitamin A-generating carotenoids in any useful quantity. However, the α- and β-carotene contents are so low that corn is never listed among the sources of vitamin A. Corn's primary carotenoids are lutein and zeaxanthin. In one study, lutein ranged from 0 to 20.0 μg/g dry weight, suggesting wide variability (Kurilich and Juvick 1999). β-Cryptoxanthin is only found in small concentrations.

Wheat, like corn, showed significant differences in terms of the concentration of various carotenoids. Lutein was the primary carotenoid present in soft wheat samples grown in Maryland, at levels of 0.82–1.14 μg/g. There were also significant amounts of zeaxanthin and β-carotene (Moore Morales de Leon *et al.* 2005). In a study of Canadian wheats, einkorn (*Triticum monococcum*) had the highest level of all-*trans*-lutein, averaging 7.41 μg/g, with small amounts of all-*trans*-zeaxanthin, *cis*-lutein isomers and β-carotene. Durum, Kamut, and Khorasan (*Triticum turgidum*) had intermediate levels of lutein (5.41–5.77 μg/g), while common bread or pastry wheat (*Triticum aestivum*) had the lowest content (2.01–2.11 μg/g). The lutein content of einkorn appeared to be influenced significantly by environmental growing conditions (Abdel-Aal *et al.* 2007).

As is true with the other phytochemicals, carotenoids can be affected by processing. The carotenoid content of whole grain breads decreased during fermentation, kneading and baking. Losses might be as great as 80%, due to enzymatic activity in the whole grain (Leenhardt *et al.* 2006).

Carotenoids—both vitamin A and non-vitamin A producing—contribute to the overall antioxidant capacity of the grain and may bolster the disease risk reduction potential of whole grains. For example, the intake of non-vitamin A-producing carotenoid, β-cryptoxanthin, was significantly associated with a reduced risk of type 2 diabetes (relative risk = 0.58; Montanen *et al.* 2003).

Lutein or zeaxanthin intake has been suggested as reducing the risk of age-related macular degeneration or cataracts. However, in 2006 FDA analysis did not find the association strong enough to warrant a health claim (Trumbo and Elwood 2006). Furthermore, levels of carotenoid suggested for the health claim are very much higher than could be achieved through normal grain ingestion. No requirement has been set for antioxidants as a group or carotenoids as a category.

Epidemiological data suggests that those with high intakes of carotenoids from food (yellow, orange and dark green vegetables and fruit and whole grains) rather than supplements, have lower disease and cancer risks. Current dietary recommendations suggest increasing intakes of these foods as a way to attain these phytonutrients.

7. Conclusions

Whole grains contain a number of functional components including long-recognised traditional nutrient components and some newly-identified phytochemicals. These components suggest numerous possible mechanisms in which whole grains could reduce the risk of developing a variety of chronic diseases. It needs to be determined whether disease reduction is due to a synergy of components provided by the whole grain package rather than any one component in isolation.

In sum, whole grains seem to offer the same net health benefits. However, different species, varieties and cultivar types do not present the same levels and types of components. Thus while two different whole grains may lower cholesterol or reduce cancer risk, they may use entirely different mechanisms.

That being said, it is important that nutrition advice emphasises the selection of a wide variety of grains in order to benefit from the unique compounds offered by each grain type. Brown rice offers oryzanol, blue corn and wheat offers some unusual anthocyanins, barley and oats contribute β-glucan, wheat contributes arabinoxylans and rye contributes high levels of alkylresorcinols. The diet benefits from all of these compounds and would benefit from adequate and varied whole grain intake.

Unfortunately, the per capita consumption of whole grain foods in the US and other developed nations falls well below the recommended 3 ounce (90 g)-equivalent servings per day. Health professionals and the public must be apprised of the benefits of including these important foods and the mother lode of functional components.

8. References

Abdel-Aal, el-S.M. and Hucl, P. 2003. Composition and stability of anthocyanins in blue-grained wheat. *Journal of Agricultural and Food Chemistry* **51**: 2174-2180.

Abdel-Aal, el-S.M., Young, J.C. and Rabalski, I. 2006. Anthocyanin composition in black, blue, pink, purple, and red cereal grains. *Journal of Agricultural and Food Chemistry* **4**: 4696-4704.

Abdel-Aal, el-S.M., Young, J.C., Rabalski, I., Hucl, P. and Fregeau-Reid, J. 2007. Identification and quantification of seed carotenoids in selected wheat species. *Journal of Agricultural and Food Chemistry* **55**: 787-794.

AACC International. 1999. Definition of whole grain. AACC International. Available at: www.aaccnet.org/definitions/wholegrain.asp.

AACC International. 2001. Definition of dietary fibre. AACC International. Available at: www.aaccnet.org/definitions/default.asp.

AACC International. 2007. Response to Health Canada regarding whole grains. Available at: www.aaccnet.org/definitions/pdfs/AACCIntlResponseToHealthCanadaFeb07.pdf.

Adlercreutz, H. 1990. Western diet and Western diseases: some hormonal and biochemical mechanisms and associations. *Scandinavian Journal of Clinical Laboratory Investigation (Supplement)* **201**: 3-23.

Adom, K.K., Sorrells, M.E. and Liu, R.H. 2005. Phytochemicals and antioxidant activity of milled fractions of different wheat varieties. *Journal of Agricultural and Food Chemistry* **53**: 2297-306.

Akihisa, T., Yasukawa, K. and Yamaura, M., *et al.* 2000. Triterpene alcohol and sterol ferulates from rice bran and their anti-inflammatory effects. *Journal of Agricultural and Food Chemistry* **48**: 2313-2319.

Alfthan, G., Tapani, K., Nissinen, K., Saarela, J. and Aro, A. 2004. The effect of low doses of betaine on plasma homocysteine in healthy volunteers. *British Journal of Nutrition* **92**: 665-669.

Alonso, A., Beunza, J.J., Bes-Rastrollo, M., Pajares, R.M. and Martinez-Gonzalez, M.A. 2006. Vegetable protein and fibre from cereal are inversely associated with the risk of hypertension in a Spanish cohort. *Archives of Medical Research* **37**: 778-786.

Alvarez, P., Alvarado, C., Mathieu, F., Jiménez, L. and De la Fuente, M. 2006a. Diet supplementation for 5 weeks with polyphenol-rich cereals improves several functions and the redox state of mouse leucocytes. *European Journal of Nutrition* **45**: 428-438.

Alvarez, P., Alvarado, C., Puerto, M., Schlumberger, A., Jiménez, L. and De la Fuente, M. 2006b. Improvement of leukocyte functions in prematurely aging mice after five weeks of diet supplementation with polyphenol-rich cereals. *Nutrition* **22**: 913-921.

Andersson, S.W., Skinner, J., Ellegard, L., *et al.* 2004. Intake of dietary plant sterols is inversely related to serum cholesterol concentration in men and women in the EPIC Norfolk population: a cross-sectional study. *European Journal of Clinical Nutrition* **58**: 1378-1385.

Barton, B.A., Eldridge, A.L., Thompson, D., *et al.* 2005. The relationship of breakfast and cereal consumption to nutrient intake and body mass index: the National Heart, Lung, and Blood Institute Growth and Health Study. *Journal of the American Dietetic Association* **105**: 1383-1389.

Bach Knudsen, K.E., Serena, A., Kjaer, A.K., *et al.* 2003. Rye bread in the diet of pigs enhances the formation of enterolactone and increases its levels in plasma, urine and feces. *Journal of Nutrition* **133**: 1368-1375.

Bach Kristensen, M., Tetens, I., Alstrup Jorgensen, A.B., *et al.* 2005. A decrease in iron status in young healthy women after long-term daily consumption of the recommended intake of fibre-rich wheat bread. *European Journal of Nutrition* **44**(6): 334-40.

Baublis, A.J., Lu, C., Clydesdale, F.M. and Decker, E.A. 2000. Potential of wheat-based breakfast cereals as a source of dietary antioxidants. *Journal of the American College of Nutrition* **19**: 308-311.

Bazzano, L.A., Song, Y., Bubes, V., Good, C.K., Manson, J.E. and Liu, S. 2005. Dietary intake of whole and refined grain breakfast cereals and weight gain in men. *Obesity Research* **13**: 1952-1960.

Behall, K.M., Scholfield, D.J., Hallfrisch, J.G. and Liljeberg-Elmstahl, H.G. 2006. Consumption of both resistant starch and β-glucan improves postprandial plasma glucose and insulin in women. *Diabetes Care* **29**: 976-981.

Berger, A., Gremaud, G., Baumgartner, M., *et al.* 2003. Cholesterol-lowering properties of amaranth grain and oil in hamsters. *International Journal of Vitaminology and Nutrition Research* **73**: 39-47.

Bering, S., Suchdev, S., Sjøltov, L., Berggren, A., Tetens, I. and Bukhave, K. 2006. A lactic acid-fermented oat gruel increases non-haem iron absorption from a phytate-rich meal in healthy women of childbearing age. *British Journal of Nutrition* **96**: 80-85.

Bingham, S.A., Day, N.E., Luben, R., *et al.* European Prospective Investigation into Cancer and Nutrition. 2003. Dietary fibre in food and protection against colorectal cancer in the European Prospective Investigation into Cancer and Nutrition (EPIC): an observational study. *Lancet* **361**: 1496-1501.

Bressani, R., Turcios, J.C., Colmenares de Ruiz, A.S. and de Palomo, P.P. 2004. Effect of processing conditions on phytic acid, calcium, iron, and zinc contents of lime-cooked maize. *Journal of Agricultural and Food Chemistry* **52**: 1157-1162.

Broadley, M.R., White, P.J., Bryson, R.J., *et al.* 2006. Biofortification of UK food crops with selenium. *Proceedings of the Nutrition Society* **65**: 169-181.

Bryngelsson, S., Dimberg, L.H. and Kamal-Eldin, A. 2002. Effects of commercial processing on levels of antioxidants in oats (*Avena sativa* L.). *Journal of Agricultural and Food Chemistry* **50**: 1890-1896.

Buyken, A.E., Toeller, M., Heitkamp, G., *et al.* 1998. Relation of fibre intake to HbA1c and the prevalence of severe ketoacidosis and severe hypoglycaemia. EURODIAB IDDM Complications Study Group. *Diabetologia* **41**: 882-890.

Campos, F.G., Logullo Waitzberg, A.G., Kiss, D.R., *et al.* 2005. Diet and colorectal cancer: current evidence for etiology and prevention. *Nutrición Hospitalaria* **20**: 18-25.

Chen, H.L., Haack, V.S., Janecky, C.W., Vollendorf, N.W. and Marlett, J.A. 1998. Mechanism by which wheat bran and oat bran increase stool weight in humans. *American Journal of Clinical Nutrition* **68**: 711-719.

Cicero, A.F. and Gaddi, A. 2001. Rice bran oil and gamma-oryzanol in the treatment of hyperlipoproteinaemias and other conditions. *Phytotherapeutic Research* **15**: 277-289.

Cleveland, L.E., Moshfegh, A.J., Albertson, A.M. and Goldman, J.D. 2000. Dietary intake of whole grains. *Journal of the American College of Nutrition* **19**(Suppl. 3): 331S-338S.

Clifford, M.N. 2004. Diet-derived phenols in plasma and tissues and their implications for health. *Planta Medica* **70**: 1103-1114.

Clifton, P.M., Noakes, M., Ross, D., Fassoulakis, A., Cehun, M. and Nestel, P. 2004. High dietary intake of phytosterol esters decreases carotenoids and increases plasma plant sterol levels with no additional cholesterol lowering. *Journal of Lipid Research* **45**: 1493-1499.

Clydesdale, F.M., Ho, C.T., Lee, C.Y., Mondy, N.I. and Shewfelt, R.L. 1991. The effects of postharvest treatment and chemical interactions on the bioavailability of ascorbic acid, thiamin, vitamin A, carotenoids, and minerals. *Critical Reviews in Food Science and Nutrition* **30**: 599-638.

Cotterchio, M., Boucher, B.A., Manno, M., Gallinger, S., Okey, A. and Harper, P. 2006. Dietary phytoestrogen intake is associated with reduced colorectal cancer risk. *Journal of Nutrition* **136**: 3046-3053.

Craig, S.A. 2004. Betaine in human nutrition. *American Journal of Clinical Nutrition* **80**: 539-549.

Curhan, G.C., Willett, W.C., Knight, E.L. and Stampfer, M.J. 2004. Dietary factors and the risk of incident kidney stones in younger women: Nurses' Health Study II. *Archives of Internal Medicine* **164**: 885-891.

Ellegard, L. and Andersson, H. 2007. Oat bran rapidly increases bile acid excretion and bile acid synthesis: an ileostomy study. *European Journal of Clinical Nutrition* **61**(8): 938-945.

Erkkila, A.T., Herrington, D.M., Mozaffarian, D. and Lichtenstein, A.H. 2005. Cereal fibre and whole-grain intake are associated with reduced progression of coronary-artery atherosclerosis in postmenopausal women with coronary artery disease. *American Heart Journal* **150**: 94-101.

Ehrenbergerova, J., Belcrediova, N., Pryma, J., Vaculova, K. and Newman, C.W. 2006. Effect of cultivar, year grown, and cropping system on the content of tocopherols and tocotrienols in grains of hulled and hulless barley. *Plant Foods for Human Nutrition* **61**: 145-150.

Ferguson, L.R., Chavan, R.R. and Harris, P.J. 2001. Changing concepts of dietary fibre: implications for carcinogenesis. *Nutrition and Cancer* **39**: 155-169.

Fischer, L.M., daCosta, K.A., Kwock, L., *et al.* 2007. Sex and menopausal status influence human dietary requirements for the nutrient choline. *American Journal of Clinical Nutrition* **85**: 1275-1285.

Fleming, D.J., Tucker, K.L., Jacques, P.F., Dallal, G.E., Wilson, P.W. and Wood, R.J. 2002. Dietary factors associated with the risk of high iron stores in the elderly Framingham Heart Study cohort. *American Journal of Clinical Nutrition* **76**(6): 1375-84.

Fox, C.H. and Eberl, M. 2002. Phytic acid (IP6), novel broad spectrum anti-neoplastic agent: a systematic review. *Complementary and Therapeutic Medicine* **10**: 229-234.

Gao, X., Wilde, P.E., Lichtenstein, A.H., Bermudez, O.I. and Tucker, K.L. 2006. The maximal amount of dietary alpha-tocopherol intake in U.S. adults (NHANES 2001–2002). *Journal of Nutrition* **136**: 1021-1026.

Garcia, A.L., Otto, B., Reich, S.C., *et al.* 2006. Arabinoxylan consumption decreases postprandial serum glucose, serum insulin and plasma total ghrelin response in subjects with impaired glucose tolerance. *European Journal of Clinical Nutrition* **61**: 334-341.

Glahn, R.P., Cheng, Z., Welch, R.M. and Gregorio, G.B. 2002. Comparison of iron bioavailability from 15 rice genotypes: studies using an *in vitro* digestion/caco-2 cell culture model. *Journal of Agricultural and Food Chemistry* **50**: 3586-3591.

Grasten, S.M., Juntunen, K.S., Poutanen, K.S., Gylling, H.K., Miettinen, T.A. and Mykkanen, H.M. 2000. Rye bread improves bowel function and decreases the concentrations of some compounds that are putative colon cancer risk markers in middle-aged women and men. *Journal of Nutrition* **130**: 2215-2221.

Hakala, P., Lampi, A.M., Ollilainen, V., *et al.* 2002. Steryl phenolic acid esters in cereals and their milling fractions. *Journal of Agricultural and Food Chemistry* **50**: 5300-5307.

Hallikainen, M., Toppinen, L., Mykkanen, H., *et al.* 2006. Interaction between cholesterol and glucose metabolism during dietary carbohydrate modification in subjects with the metabolic syndrome. *American Journal of Clinical Nutrition* **84**: 1385-1392.

Hambraeus, L. 1999. Animal- and plant-food-based diets and iron status: benefits and costs. *Proceedings of the Nutrition Society* **58**: 235-242.

Hargrove, J.L., Greenspan, P. and Hartle, D.K. 2004. Nutritional significance and metabolism of very long chain fatty alcohols and acids from dietary waxes. *Experimental Biology and Medicine* (Maywood) **229**: 215-226.

He, K., Liu, K. and Daviglus, M.L., *et al.* 2006. Magnesium intake and incidence of metabolic syndrome among young adults. *Circulation* **113**: 1675-1682.

Hemalatha, S., Platel, K. and Srinivasan, K. 2007. Influence of heat processing on the bioaccessibility of zinc and iron from cereals and pulses consumed in India. *Journal of Trace Elements in Medicine and Biology* **21**: 1-7.

Hietaniemi, V. and Poutanen, K. 2003. Process-induced changes on bioactive compounds in whole grain rye. *Proceedings of the Nutrition Society* **62**: 117-122.

Hung, T., Sievenpiper, J.L., Marchie, A., Kendall, C.W. and Jenkins, D.J. 2003. Fat versus carbohydrate in insulin resistance, obesity, diabetes and cardiovascular disease. *Current Opinions in Clinical Nutrition and Metabolic Care* **6**: 165-176.

Hunt, J.R. 2002. Moving toward a plant-based diet: are iron and zinc at risk? *Nurition Reviews* **60**: 127-134.

Hurrell, R.F. 2003. Influence of vegetable protein sources on trace element and mineral bioavailability. *Journal of Nutrition* **133**: 2973S-2977S.

Irmak, S. and Dunford, N.T. 2005. Policosanol contents and compositions of wheat varieties. *Journal of Agricultural and Food Chemistry* **53**: 5583-5586.

Iwatsuki, K., Akihisa, T. and Tokuda, H., *et al.* 2003. Sterol ferulates, sterols, and 5-alk(en)ylresorcinols from wheat, rye, and corn bran oils and their inhibitory effects on Epstein-Barr virus activation. *Journal of Agricultural and Food Chemistry* **51**: 6683-6688.

Jariwalla, R.J. 1999. Inositol hexaphosphate (IP6) as an anti-neoplastic and lipid-lowering agent. *Anticancer Research* **19**(5A): 3699-3702.

Jariwalla, R.J. 2001. Rice-bran products: phytonutrients with potential applications in preventive and clinical medicine. *Drugs in Experimental and Clinical Research* **27**: 17-26.

Jenab, M. and Thompson, L.U. 2000. Phytic acid in wheat bran affects colon morphology, cell differentiation and apoptosis. *Carcinogenesis* **21**: 1547-1552.

Jenkins, A.L., Jenkins, D.J., Zdravkovic, U., Wursch, P. and Vuksan, V. 2002. Depression of the glycemic index by high levels of beta-glucan fibre in two functional foods tested in type 2 diabetes. *European Journal of Clinical Nutrition* **56**: 622-628.

Jensen, M.K., Koh-Banerjee, P. and Hu, F.B., *et al.* 2004. Intakes of whole grains, bran, and germ and the risk of coronary heart disease in men. *American Journal of Clinical Nutrition* **80**: 1492-1499.

Jones, J.M. 2004. Dietary fibre intake, disease prevention, and health promotion. In: van der Kamp, J.W., Asp, N-G., Miller Jones, J. and Schaafsma, G. *Dietary fibre: bio-active carbohydrates for food and feed*: 143-165. Wageningen Academic, The Netherlands.

Juntunen, K.S., Mazur, W.M. and Liukkonen, K.H., *et al.* 2000. Consumption of wholemeal rye bread increases serum concentrations and urinary excretion of enterolactone compared with consumption of white wheat bread in healthy Finnish men and women. *British Journal of Nutrition* **84**: 839-846.

Kanaya, Y., Doi, T. and Sasaki, H., *et al.* 2004. Rice bran extract prevents the elevation of plasma peroxylipid in KKAy diabetic mice. *Diabetes Research and Clinical Practice* **66**(Suppl. 1): S157-S160.

Karmally, W., Montez, M.G. and Palmas, W., *et al.* 2005. Cholesterol-lowering benefits of oat-containing cereal in Hispanic Americans. *Journal of the American Dietetics Association* **105**: 967-970.

Katz, D.L., Nawaz, H. and Boukhalil, J., *et al.* 2003. Effects of oat and wheat cereals on endothelial responses. *Preventive Medicine* **33**: 476-484.

Kaukovirta-Norja, A. 2006. *Barley and rye – new potential for healthy breakfast cereals/foods.* AACC International World Grains Summit and Annual Meeting, San Francisco, USA.

Keenan, J.M., Pins, J.J., Frazel., C., Moran, A. and Turnquist, L. 2002. Oat ingestion reduces systolic and diastolic blood pressure in patients with mild or borderline hypertension: a pilot trial. *Journal of Family Practice* **51**: 369.

Kendall, C.W., Emam, A., Augustin, L.S. and Jenkins, D.J. 2004. Resistant starches and health. *Journal of American Association of Analytical Chemists International* **87**: 769-774.

Kerckhoffs, D.A., Brouns, F., Hornstra, G. and Mensink, R.P. 2002. Effects on the human serum lipoprotein profile of beta-glucan, soy protein and isoflavones, plant sterols and stanols, garlic and tocotrienols. *Journal of Nutrition* **132**: 2494-2505.

Klingberg, S., Andersson, H. and Mulligan, A., *et al*. 2007. Food sources of plant sterols in the EPIC Norfolk population. *European Journal of Clinical Nutrition*. Advance online publication 18 April 2007; doi: 10.1038/sj.ejcn.1602765.

Koh-Banerjee, P., Franz, M. and Sampson, L., *et al*. 2004. Changes in whole-grain, bran, and cereal fibre consumption in relation to 8-y weight gain among men. *American Journal Clinical Nutrition* **80**: 1237-1245.

Krajcovicova-Kudlackova, M., Babinska, K. and Valachovicova, M. 2005. Health benefits and risks of plant proteins. *Bratislavské Lekárske Listy* **106**: 231-234.

Kreijkamp-Kaspers, S., Kok, L., Bots, M.L., Grobbee, D.E. and van der Schouw, Y.T. 2005. Dietary phytoestrogens and plasma lipids in Dutch postmenopausal women: a cross-sectional study. *Atherosclerosis* **178**: 95-100.

Kristensen, M.B., Hels, O., Morberg, C.M., Marving, J., Bugel, S and Tetens, I. 2006. Total zinc absorption in young women, but not fractional zinc absorption, differs between vegetarian and meat-based diets with equal phytic acid content. *British Journal of Nutrition* **95**: 963-967.

Krittaphol, W., Bailey, K.B., Pongcharoen, T., Winichagoon, P. and Gibson, R.S. 2006. Low zinc, iron, and calcium intakes of Northeast Thai school children consuming glutinous rice-based diets are not exacerbated by high phytate. *International Journal of Food Science and Nutrition* **57**: 520-528.

Kurilich, A.C. and Juvik, J.A. 1999. Quantification of carotenoid and tocopherol antioxidants in Zea mays. *Journal of Agricultural and Food Chemistry* **47**: 1948-1955.

Lairon, D., Arnault, N., Bertrais, S., Planells, R., Clero, E., Hercberg, S. and Boutron-Ruault, M.C. 2005. Dietary fibre intake and risk factors for cardiovascular disease in French adults. *American Journal of Clinical Nutrition* **82**: 1185-1194.

La Vecchia, C., Chatenoud, L., Negri, E. and Franceschi, S. 2003. Session: whole cereal grains, fibre and human cancer wholegrain cereals and cancer in Italy. *Proceedings of the Nutrition Society* **62**: 45-49.

Leenhardt, F., Lyan, B., Rock, E., Boussard, A., Potus, J., Chanliaud, E. and Remesy, C. 2006. Wheat lipoxygenase activity induces greater loss of carotenoids than vitamin E during breadmaking. *Journal of Agricultural and Food Chemistry* **54**: 1710-1715.

Lestienne, I., Caporiccio, B., Besancon, P., Rochette, I. and Treche, S. 2005. Relative contribution of phytates, fibres, and tannins to low iron and zinc *in vitro* solubility in pearl millet (*Pennisetum glaucum*) flour and grain fractions. *Journal of Agricultural and Food Chemistry* **53**: 8342-8348.

Levrat-Verny, M.A., Coudray, C., Bellanger, J., Lopez, H.W., Demigne, C., Rayssiguier, Y. and Remesy, C. 1999. Whole-wheat flour ensures higher mineral absorption and bioavailability than white wheat flour in rats. *British Journal of Nutrition* **82**: 17-21.

Lewis, J. and Buss, D.H. 1988. Trace nutrients: 5. Minerals and vitamins in the British household food supply. *British Journal of Nutrition* **60**: 413-424.

Liese, A.D., Roach, A.K., Sparks, K.C., Marquart, L., D'Agostino, R.B. and Mayer-Davis, E.J. 2003. Whole-grain intake and insulin sensitivity: the Insulin Resistance Atherosclerosis Study. *American Journal of Clinical Nutrition* **78**: 965-971.

Lim, C.C., Ferguson, L.R. and Tannock, G.W. 2005. Dietary fibres as "prebiotics": implications for colorectal cancer. *Molecular Nutrition and Food Research* **49**: 609-619.

Linseisen, J., Piller, R., Hermann, S. and Chang-Claude, J. 2004. Dietary phytoestrogen intake and premenopausal breast cancer risk in a German case-control study. *International Journal of Cancer* **110**: 284-290.

Liu, S., Manson, J.E. and Stampfer, M.J., *et al*. 2000a. Whole grain consumption and risk of ischemic stroke in women: a prospective study. *Journal of the American Medical Association* **284**: 1534-1540.

Liu, S., Manson, J.E., Stampfer, M.J., Hu, F.B., Giovannucci, E., Colditz, G.A., Hennekens, C.H. and Willett, W.C. 2000b. A prospective study of whole-grain intake and risk of type 2 diabetes mellitus in US women. *American Journal of Public Health* **90**: 1409-1415.

Liukkonen, K.H., Katina, K. and Wilhelmsson, A., *et al*. 2003. Process-induced changes on bioactive compounds in whole grain rye. *Proceedings of the Nutrition Society* **62**: 117-122.

Lopez, H.W., Levrat, M.A., Guy, C., Messager, A., Demigne, C. and Remesy, C. 1999. Effects of soluble corn bran arabinoxylans on cecal digestion, lipid metabolism, and mineral balance (Ca, Mg) in rats. *Journal of Nutritional Biochemistry* **10**: 500-509.

Lopez, H.W., Duclos, V., Coudray, C., *et al*. 2003. Making bread with sourdough improves mineral bioavailability from reconstituted whole wheat flour in rats. *Nutrition* **19**: 524-530.

Luo, H.F., Li, Q., Yu, S., Badger, T.M. and Fang, N. 2005. Cytotoxic hydroxylated triterpene alcohol ferulates from rice bran. *Journal of Natural Products* **68**: 94-97.

Lundin, E.A., Zhang, J.X. and Lairon, D., *et al*. 2004. Effects of meal frequency and high-fibre rye-bread diet on glucose and lipid metabolism and ileal excretion of energy and sterols in ileostomy subjects. *European Journal of Clinical Nutrition* **58**: 1410-1419.

Lyons, G.H, Judson, G.J., Ortiz-Monasterio, I., Genc, Y., Stangoulis, J.C. and Graham, R.D. 2005. Selenium in Australia: selenium status and biofortification of wheat for better health. *Journal of Trace Elements in Medicine and Biology* **19**: 75-82.

Maki, K.C., Galant, R. and Samuel, P., *et al*. 2007. Effects of consuming foods containing oat beta-glucan on blood pressure, carbohydrate metabolism and biomarkers of oxidative stress in men and women with elevated blood pressure. *European Journal of Clinical Nutrition* **61**: 786-95.

Manach, C., Scalbert, A., Morand, C., Remesy, C. and Jimenez, L. 2004. Polyphenols: food sources and bioavailability. *American Journal of Clinical Nutrition* **79**: 727-747.

Martirosyan, D.M., Miroshnichenko, L.A., Kulakova, S.N., Pogojeva, A.V. and Zoloedov, V.I. 2007. Amaranth oil application for coronary heart disease and hypertension. *Lipids in Health Disease* **6**: 1.

McCann, S.E., Kulkarni, S. and Trevisan, M., *et al*. 2006. Dietary lignan intakes and risk of breast cancer by tumor estrogen receptor status. *Breast Cancer Research and Treatment* **99**: 309-311.

McCann, S.E., Muti, P., Vito, D., Edge, S.B., Trevisan, M. and Freudenheim, J.L. 2004. Dietary lignan intakes and risk of pre- and postmenopausal breast cancer. *International Journal of Cancer* **111**: 440-443.

McCarty, M.F. 2005. Magnesium may mediate the favorable impact of whole grains on insulin sensitivity by acting as a mild calcium antagonist. *Medical Hypotheses* **64**: 619-627.

McCarty, M.F. 2006. Toward prevention of Alzheimers disease—potential nutraceutical strategies for suppressing the production of amyloid beta peptides. *Medical Hypotheses* **67**: 682-697.

Merrill, A.H., Jr, Schmelz, E.M., Wang, E., Schroeder, J.J., Dillehay, D.L. and Riley, R.T. 1995. Role of dietary sphingolipids and inhibitors of sphingolipid metabolism in cancer and other diseases. *Journal of Nutrition* **125**(Suppl. 6): 1677S-1682S.

Menendez, R., Mas, R. and Amor, A.M., *et al.* 2000. Effects of policosanol treatment on the susceptibility of low density lipoprotein (LDL) isolated from healthy volunteers to oxidative modification *in vitro*. *British Journal of Clinical Pharmacology* **50**: 255-262.

Merchant, A.T., Hu, F.B., Spiegelman, D., Willett, W.C., Rimm, E.B. and Ascherio, A. 2003. Dietary fibre reduces peripheral arterial disease risk in men. *Journal of Nutrition* **133**: 3658-3663.

Meyer, K.A., Kushi, L.H., Jacobs, D.R. Jr., Slavin, J., Sellers, T.A. and Folsom, A.R. 2000. Carbohydrates, dietary fibre, and incident type 2 diabetes in older women. *American Journal of Clinical Nutrition* **71**: 921-930.

Milder, I.E., Feskens, E.J., Arts, I.C., Bueno-de-Mesquita, H.B., Hollman, P.C. and Kromhout, D. 2006. Intakes of 4 dietary lignans and cause-specific and all-cause mortality in the Zutphen Elderly Study. *American Journal of Clinical Nutrition* **84**: 400-405.

Miller, H.E., Rigelhof, F., Marquart, L., Prakash, A. and Kanter, M. 2000. Antioxidant content of whole grain breakfast cereals, fruits and vegetables. *Journal of the American College of Nutrition* **19**(Suppl. 3): 312S-319S.

Mirkin, A., Mas, R. and Martinto, M., *et al.* 2001. Efficacy and tolerability of policosanol in hypercholesterolemic postmenopausal women. *International Journal of Clinical Pharmacology Research* **21**: 31-41.

Montonen, J., Knekt, P., Jarvinen, R., Aromaa, A. and Reunanen, A. 2003. Whole-grain and fibre intake and the incidence of type 2 diabetes. *American Journal of Clinical Nutrition* **77**: 622-629; comment 527-529.

Montonen, J., Knekt, P., Jarvinen, R. and Reunanen, A. 2004. Dietary antioxidant intake and risk of type 2 diabetes. *Diabetes Care* **27**: 362-366.

Moore Morales de Leon, J., Camacho, M.E. and Bourges, H. 2005. Amino acid composition of some Mexican foods. *Archivos of Latinoamericanos de Nutrición* **55**: 172-186.

Moreau, R.A., Singh, V., Nunez, A. and Hicks, K.B. 2000. Phytosterols in the aleurone layer of corn kernels. *Biochemistry Society Transactions* **28**: 803-806.

Moreau, R.A. and Hicks, K.B. 2006. Reinvestigation of the effect of heat pretreatment of corn fibre and corn germ on the levels of extractable tocopherols and tocotrienols. *Journal of Agricultural and Food Chemistry* **54**: 8093-8102.

Murakami, K., Okubo, H. and Sasaki, S. 2005. Effect of dietary factors on incidence of type 2 diabetes: a systematic review of cohort studies. *Journal of Nutrition Science and Vitaminology (Tokyo)* **51**: 292-310.

Nguyen, K.A., Cao, Y., Chen, J.R., Townsend, C.M. Jr. and Ko, T.C. 2006. Dietary fibre enhances a tumor suppressor signaling pathway in the gut. *Annals of Surgery* **243**: 619-625; discussion 625-627.

Nielsen, B.K., Thompson, L.U. and Bird, R.P. 1987. Effect of phytic acid on colonic epithelial cell proliferation. *Cancer Letters* **37**: 317-325.

Nielsen, M.M. and Hansen, A. 2007. The content of vitamin E in organic wheat and spelt flour depending on the type of grinding. Cereals in Europe. AACC International European Section meeting. 5 May, 2007.

Nilsson, A. and Duan, R.D. 2006. Absorption and lipoprotein transport of sphingomyelin. *Journal of Lipid Research* **47**: 154-171.

Olthof, M.R., van Vliet, T., Verhoef, P, Zock, P.L. and Katan, M.B. 2005. Effect of homocysteine-lowering nutrients on blood lipids: results from four randomised, placebo-controlled studies in healthy humans. *PLoS Medicine* **2**: e135.

Ortega, L.L., Sanchez, J. and Mas, R., *et al.* 2006. Effects of policosanol on patients with ischemic stroke: a pilot open study. *Journal of Medical Foods* **9**: 378-385.

Ostlund, R.E. Jr, Racette, S.B., Okeke, A. and Stenson, W.F. 2002. Phytosterols that are naturally present in commercial corn oil significantly reduce cholesterol absorption in humans. *American Journal* of *Clinical Nutrition* **75**: 1000-1004.

Ostlund, R.E. Jr and Lin, X. 2006. Regulation of cholesterol absorption by phytosterols. *Current Atherosclerosis Reports* **8**: 487-491.

Owen, A.J., Roach, P.D. and Abbey, M. 2004. Regulation of low-density lipoprotein receptor activity by estrogens and phytoestrogens in a HepG2 cell model. *Annals of Nutrition and Metabolism* **48**: 269-275.

Padron, J.M. 2006. Sphingolipids in anticancer therapy. *Current Medicinal Chemistry* **13**: 755-770.

Panfili, G., Fratianni, A. and Irano, M. 2003. Normal phase high-performance liquid chromatography method for the determination of tocopherols and tocotrienols in cereals. *Journal of Agricultural and Food Chemistry* **51**: 3940-3944.

Panlasigui, L.N. and Thompson, L.U. 2006. Blood glucose lowering effects of brown rice in normal and diabetic subjects. *International Journal of Food Science and Nutrition* **57**: 151-158.

Park, Y., Hunter, D.J. and Spiegelman, D., *et al.* 2005. Dietary fibre intake and risk of colorectal cancer: a pooled analysis of prospective cohort studies. *Journal of the American Medical Association* **294**(22): 2849-2857.

Pereira, M.A., Jacobs, D.R. Jr. and Pins, J.J., *et al.* 2002. Effect of whole grains on insulin sensitivity in overweight hyperinsulinemic adults. *American Journal of Clinical Nutrition* **75**: 848-855.

Pereira, M.A., O'Reilly, E. and Augustsson, K., *et al.* 2004. Dietary fibre and risk of coronary heart disease: a pooled analysis of cohort studies. *Archives of Internal Medicine* **4**: 370-376.

Perlas, L.A. and Gibson, R.S. 2005. Household dietary strategies to enhance the content and bioavailability of iron, zinc and calcium of selected rice- and maize-based Philippine complementary foods. *Maternal and Child Nutrition* **4**: 263-273.

Pool-Zobel, B.L., Adlercreutz, H. and Glei, M., *et al.* 2000. Isoflavonoids and lignans have different potentials to modulate oxidative genetic damage in human colon cells. *Carcinogenesis* **6**: 1247-1252.

Poppitt, S.D., van Drunen, J.D., McGill, A.T., Mulvey, T.B. and Leahy, F.E. 2007. Supplementation of a high-carbohydrate breakfast with barley beta-glucan improves postprandial glycaemic response for meals but not beverages. *Asia Pacific Journal of Clinical Nutrition* **16**: 16-24.

Proulx, A.K. and Reddy, M.B. 2007. Fermentation and lactic acid addition enhance iron bioavailability of maize. *Journal of Agricultural and Food Chemistry* **55**: 2749-2754.

Qureshi, A.A., Mo, H., Packer, L. and Peterson, D.M. 2000. Isolation and identification of novel tocotrienols from rice bran with hypocholesterolemic, antioxidant, and antitumor properties. *Journal of Agricultural and Food Chemistry* **48**: 3130-3140.

Qureshi, A.A., Sami, S.A., Salser, W.A. and Khan, F.A. 2002. Dose-dependent suppression of serum cholesterol by tocotrienol-rich fraction (TRF25) of rice bran in hypercholesterolemic humans. *Atherosclerosis* **161**: 199-207.

Ramakers, J.D., Volman, J.J., Biorklund, M., Onning, G., Mensink, R.P. and Plat, J. 2007. Fecal water from ileostomic patients consuming oat beta-glucan enhances immune responses in enterocytes. *Molecular Nutrition and Food Research* **51**: 211-220.

Ripsin, C.M., Keenan, J.M. and Jacobs, D.R. Jr., *et al.* 1992. Oat products and lipid lowering. A meta-analysis. *Journal of the American Medical Association* **67**: 3317-3325: Erratum in **268**: 3074.

Ross, A.B., Shepherd, M.J., Schupphaus, M., Sinclair, V., Alfaro, B., Kamal-Eldin, A. and Aman, P. 2003. Alkylresorcinols in cereals and cereal products. *Journal of Agricultural and Food Chemistry* **51**: 4111-4118.

Sandstrom, B., Almgren, A., Kivisto, B. and Cederblad, A. 1987. Zinc absorption in humans from meals based on rye, barley, oatmeal, triticale and whole wheat. *Journal of Nutrition* **117**: 1898-1902.

Scavariello, E.M. and Arellano, D.B. 1998. Gamma-oryzanol: an important component in rice brain oil. *Archivos of Latinoamericanos de Nutrición* **48**: 7-12.

Smith, T.J. 2000. Squalene: potential chemopreventive agent. *Expert Opinion in Investigational Drugs* **9**: 1841-1848.

Song, Y., Sesso, H.D., Manson, J.E., Cook, N.R., Buring, J.E. and Liu, S. 2006. Dietary magnesium intake and risk of incident hypertension among middle-aged and older US women in a 10-year follow-up study. *American Journal of Cardiology* **98**: 1616-1621.

Stattin, P., Bylund, A., Biessy, C., Kaaks, R., Hallmans, G. and Adlercreutz, H. 2004. Prospective study of plasma enterolactone and prostate cancer risk (Sweden). *Cancer Causes and Control* **15**: 1095-1102.

Stevens, J., Ahn, K., Juhaeri, Houston, D., Steffan, L. and Couper, D. 2002. Dietary fibre intake and glycemic index and incidence of diabetes in African-American and white adults: the ARIC study. *Diabetes Care* **25**: 1715-1721.

Sugawara, T. and Miyazawa, T. 1999. Separation and determination of glycolipids from edible plant sources by high-performance liquid chromatography and evaporative light-scattering detection. *Lipids* **34**: 1231-1237.

Sugawara, T., Kinoshita, M., Ohnishi, M. and Miyazawa, T. 2002. Apoptosis induction by wheat-flour sphingoid bases in DLD-1 human colon cancer cells. *Bioscience, Biotechnology and Biochemistry* **66**: 2228-2231.

Sugawara, T., Kinoshita, M., Ohnishi, M., Nagata, J. and Saito, M. 2003. Digestion of maize sphingolipids in rats and uptake of sphingadienine by Caco-2 cells. *Journal of Nutrition* **133**: 2777-2782.

Thanos, J., Cotterchio, M., Boucher, B.A., Kreiger, N. and Thompson, L.U. 2006. Adolescent dietary phytoestrogen intake and breast cancer risk (Canada). *Cancer Causes and Control* **17**: 1253-1261.

Thompson, L.U., Robb, P., Serraino, M. and Cheung, F. 1991. Mammalian lignan production from various foods. *Nutrition Cancer* **16**: 43-52.

Trumbo, P.R. and Ellwood, K.C. 2006. Lutein and zeaxanthin intakes and risk of age-related macular degeneration and cataracts: an evaluation using the Food and Drug Administration's evidence-based review system for health claims. *American Journal of Clinical Nutrition* **84**: 971-974.

Tsuji, K., Mitsutake, S. and Ishikawa, J., *et al.* 2006. Dietary glucosylceramide improves skin barrier function in hairless mice. *Journal of Dermatological Science* **44**: 101-107.

US Department of Health and Human Services., Food and Drug Administration. 2006. Food labeling: health claims; soluble dietary fibre from certain foods and coronary heart disease. Final rule. *Federal Register* **71**(98): 29248-29250.

US Dietary Guidelines Advisory Committee. 2005. Report of the Dietary Guidelines Advisory Committee on dietary guidelines for Americans. Available at: www.health.gov/dietaryguidelines/dga2005/report.

van Bennekum, A.M., Nguyen, D.V., Schulthess, G., Hauser, H. and Phillips, M.C. 2005. Mechanisms of cholesterol-lowering effects of dietary insoluble fibres: relationships with intestinal and hepatic cholesterol parameters. *British Journal of Nutrition* **94**: 331-337.

van der Schouw, Y.T., Pijpe, A. and Lebrun, C.E., *et al.* 2002. Higher usual dietary intake of phytoestrogens is associated with lower aortic stiffness in postmenopausal women. *Arteriosclerosis, Thrombosis, and Vascular Biology* **22**: 1316-1322.

van Dam, R.M., Hu, F.B., Rosenberg, L., Krishnan, S. and Palmer, J.R. 2006. Dietary calcium and magnesium, major food sources, and risk of type 2 diabetes in U.S. black women. *Diabetes Care* **29**: 2238-2243.

Vanharanta, M., Voutilainen, S., Lakka, T.A., van der Lee, M., Adlercreutz, H. and Salonen, J.T. 1999. Risk of acute coronary events according to serum concentrations of enterolactone: a prospective population-based case-control study. *Lancet* **354**: 2112-2115.

Wang, B.Y., Singer, A.H., Tsao, P.S., Drexler, H., Kosek, J. and Cooke, J.P. 1994. Dietary arginine prevents atherogenesis in the coronary artery of the hypercholesterolemic rabbit. *Journal of American College of Cardiology* **23**: 452-458.

Weickert, M.O., Mohlig, M. and Schofl, C., *et al.* 2006. Cereal fibre improves whole-body insulin sensitivity in overweight and obese women. *Diabetes Care* **29**: 775-780.

Welch, R.M., House, W.A., Ortiz-Monasterio, I. and Cheng, Z. 2005. Potential for improving bioavailable zinc in wheat grain (*Triticum* species) through plant breeding. *Journal of Agricultural and Food Chemistry* **53**: 2176-2180.

Whole Grains Council. 2003. Definition of whole grains. Available at: www.wholegrainscouncil.org/whole-grains-101/definition-of-whole-grains.

Wilson, T.A., Nicolosi, R.J., Woolfrey, B. and Kritchevsky, D. 2007. Rice bran oil and oryzanol reduce plasma lipid and lipoprotein cholesterol concentrations and aortic cholesterol ester accumulation to a greater extent than ferulic acid in hypercholesterolemic hamsters. *Journal of Nutritional Biochemistry* **18**: 105-112.

Wong, J.M., de Souza, R., Kendall, C.W., Emam, A. and Jenkins, D.J. 2006. Colonic health: fermentation and short chain fatty acids. *Journal of Clinical Gastroenterology* **40**: 235-243.

Worgall, T.S. 2007. Sphingolipids: major regulators of lipid metabolism. *Current Opinions in Clinical Nutritional Metabolic Care* **10**: 149-155.

Wright, M.E., Weinstein, S.J., Lawson, K.A., Albanes, D, Subar, A.F., Dixon, L.B., Mouw, T., Schatzkin, A.T. and Leitzmann, M.F. 2007. Supplemental and dietary vitamin E intakes and risk of prostate cancer in a large prospective study. *Cancer Epidemiology Biomarkers and Prevention* **16**: 1128-1135.

Xu, Z., Hua, N. and Godber, J.S. 2001. Antioxidant activity of tocopherols, tocotrienols, and gamma-oryzanol components from rice bran against cholesterol oxidation accelerated by

2,2'-azobis(2-methylpropionamidine) dihydrochloride. *Journal of Agricultural and Food Chemistry* **49**: 2077-2081.

Yang, X.E., Chen, W.R. and Feng, Y. 2007. Improving human micronutrient nutrition through biofortification in the soil-plant system: China as a case study. *Environmental Geochemistry and Health* **29**: 413-428.

Ylonen, K., Saloranta, C., Kronberg-Kippila, C., Groop, L., Aro, A. and Virtanen, S.M. 2003. Associations of dietary fibre with glucose metabolism in nondiabetic relatives of subjects with type 2 diabetes: the Botnia Dietary Study. *Diabetes Care* **26**: 1979-1985.

Yonekura, L. and Suzuki, H. 2005. Effects of dietary zinc levels, phytic acid and resistant starch on zinc bioavailability in rats. *European Journal of Nutrition* **44**: 384-391.

Yoon, J.H., Thompson, L.U. and Jenkins, D.J. 1983. The effect of phytic acid on *in vitro* rate of starch digestibility and blood glucose response. *American Journal of Clinical Nutrition* **38**: 835-842.

Zeisel, S.H., Mar, M.H., Howe, J.C. and Holden, J.M. 2003. Concentrations of choline-containing compounds and betaine in common foods. *Journal of Nutrition* **133**: 1302-1207.

Zeleniuch-Jacquotte, A., Lundin, E. and Micheli, A., *et al.* 2006. Circulating enterolactone and risk of endometrial cancer. *International Journal of Cancer* **119**: 2376-2381.

Zhang, Z., Song, Y. and Wang, X.L. 2005. Inositol hexaphosphate-induced enhancement of natural killer cell activity correlates with suppression of colon carcinogenesis in rats. *World Journal of Gastroenterology* **11**: 5044-5046.

Zielinski, H., Michalska, A., Piskula, M.K. and Kozlowska, H. 2006. Antioxidants in thermally treated buckwheat groats. *Molecular Nutrition and Food Research* **50**: 824-832.

Microbial conjugated linoleic acid production - a novel probiotic trait?

Rebecca Wall[1,2,3], R. Paul Ross[1,2], Gerald F. Fitzgerald[1,3] and Catherine Stanton[1,2]*

[1]*Alimentary Pharmabiotic Centre (APC), Co. Cork, Ireland*

[2]*Teagasc, Moorepark, Fermoy, Co. Cork, Ireland. Tel. 00353 (0)2542222. Fax 00353 (0)2542340. E-mail catherine.stanton@teagasc.ie*

[3]*University College Cork, National University of Ireland, Ireland.*

**Corresponding author*

Abstract

Conjugated linoleic acid (CLA), which occurs in milk and meat as a result of microbial biohydrogenation of polyunsaturated fatty acids, has attracted much attention as a novel type of biologically beneficial functional lipid because of its numerous health benefits, which include anti-inflammatory, antidiabetic and anticarcinogenic effects. CLA formation from linoleic acid (LA) by rumen bacteria has been well documented since the early 1960s; more recently it was found that certain bacteria that naturally inhabit the human gastrointestinal tract (GIT) also possess the ability to form CLA, particularly some strains of bifidobacteria and lactobacilli as well as food-grade strains such as propionibacteria. Through the ingestion of such bacteria in functional foods there is the prospect that human CLA status could be enhanced via *in vivo* microbial formation in the intestine, thereby contributing to human health.

Keywords: conjugated linoleic acid, microbial production, gut microbiota, human health, probiotics

1. Introduction

In recent years, the health-promoting properties of dairy fats have been highlighted by several scientific investigators (Stanton *et al.* 2003; Lock and Bauman 2004; Bhattacharya *et al.* 2006). In particular, conjugated linoleic acid (CLA) has received considerable attention on account of its biological properties. CLA is a collective term for a group of positional and geometric isomers of linoleic acid (LA; *c*9, *c*12 18:2*n*-6) with a conjugated double bond system instead of isolated double bonds (Figure 1; Siems *et al.* 2001). CLA is produced through the action of anaerobic rumen bacteria when dietary linoleic and α-linolenic acids obtained from plant material are converted to vaccenic acid (*trans* 11 C18:1) in the rumen (Kepler *et al.* 1966). Consequently, the principal food sources of CLA are dairy and meat products of ruminant origin (cattle, deer, sheep and goats). Since the discovery of CLA as an anticarcinogen in the 1980s, it has attracted considerable attention and has been reported to have several beneficial effects in health-related disorders. For example, CLA has been shown to reduce the risk of heart disease and several types of cancer, to decrease body fat, and to enhance bone formation and immune function in both animal models and human cell culture studies (Kritchevsky *et al.* 2000; Bassaganya-Riera *et al.* 2002; Miller *et al.* 2003; Terpstra 2004). CLA also has reported beneficial regulatory effects on lipid and eicosanoid metabolism, and cytokine and immunoglobulin production, and can modulate the expression of a number of genes, either directly or through specific transcription factors involved in the many metabolic processes they affect (Wahle *et al.* 2004). Even though there are 28 different isomers of CLA, the reported health benefits have been attributed primarily to two of its isomers: *cis*-9, *trans*-11 CLA (*c*9, *t*11 18:2) and *trans*-10, *cis*-12 CLA (*t*10, *c*12 18:2; Pariza *et al.* 2001).

Given the range of health-promoting activities attributed to CLA, interest in the development of CLA-enriched foods has increased markedly in recent years. A range of different approaches have been developed to enrich foods—especially those of dairy origin—with CLA, including different animal feeding strategies and the use of starter cultures with high CLA-producing potential in the

Food Science and Technology Bulletin: Functional Foods **4** (8) 87–97
DOI: 10.1616/1476-2137.15194. Published 28 March 2008

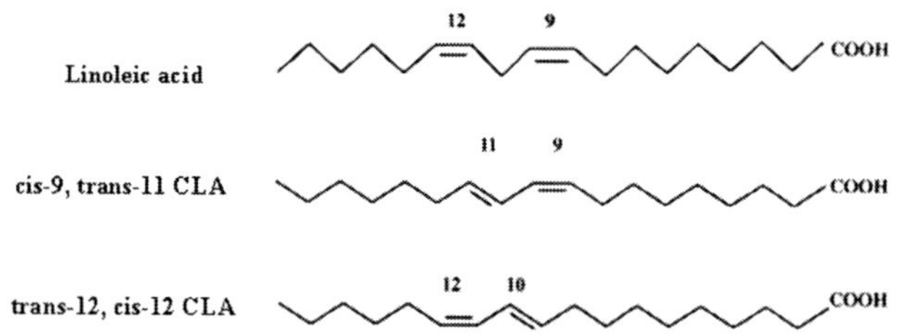

Figure 1. Structures of parent omega-6 fatty acid linoleic acid and its two main conjugated derivatives.

production of fermented dairy products. In addition, bacteria that are natural inhabitants of the human gastrointestinal tract (GIT), such as *Lactobacillus* and *Bifidobacterium*, have recently been shown to possess the ability to produce CLA and may therefore serve as a natural mechanism for promoting gut health. Some bacteria within the gastrointestinal microbiota can produce a range of other bioactive fatty molecules, in addition to CLA, such as short-chain fatty acids (acetate, propionate and butyrate) and γ-aminobutyric acid (GABA), which have been shown to possess several therapeutic properties (Topping and Clifton 2001; Hinnebusch *et al.* 2002; Coxam 2005). Since all of these microbially-produced compounds demonstrate great potential in the treatment of lifestyle diseases such as obesity, diabetes, cancer and cardiovascular disease, this makes these bacteria an important health asset for the human body. The purpose of this review is to provide an overview of CLA with a focus on its microbial production by human intestinally-derived bacteria and the potential role for this in health and disease.

2. Origin of CLA in milk and meat

CLA is formed as an intermediate in the biohydrogenation of linoleic acid via the activity of linoleic acid isomerase enzyme of ruminant bacteria, such as *Butyrivibrio fibrisolvens* (Kepler *et al.* 1966), and for several decades ruminal production was believed to be the primary source of CLA *in vivo*. As a consequence, a common name—rumenic acid—was proposed for the *c*9, *t*11 CLA isomer on the basis of its ruminal origin. However, recent findings illustrate that, although ruminal production is the source of a portion of *c*9, *t*11 CLA in ruminant tissues and milk, the majority (i.e. 70–95%) is produced through endogenous synthesis via Δ^9-desaturase enzyme from the substrate, ruminally-derived vaccenic acid (Griinari *et al.* 2000; Kay *et al.* 2004). As a result of biohydrogenation and desaturation reactions in ruminants, food products from these animals, particularly dairy products, are the major dietary sources of CLA for humans. CLA concentrations in dairy products typically range from approximately 2.8 to 7 mg/g fat, of which the *c*9, *t*11 isomer makes up 73–93% of the total CLA (Table 1; MacDonald 2000). Measurements of *c*9, *t*11 CLA in human adipose tissue indicate that its presence is highly correlated with milk fat intake (Jiang *et al.* 1999). CLA intake averages about 200 mg/day for men and 150 mg/day for women (Ritzenthaler *et al.* 2001). Several factors influence the CLA content of food products, such as animal diet, age, breed and seasonal factors. Evidence suggests that CLA increases linearly when animals are pasture–fed and decreases when grass intake declines (Stanton *et al.* 1997; French *et al.* 2000). Typically, cow's milk produced from late spring through to early fall (seasons of rapidly growing green grass) will have almost twice the CLA content compared to milk produced during winter months.

Table 1. CLA content of various foods

Food	mg/g fat	Food	mg/g fat
Dairy products		Meats/Fish	
Homogenised milk	5.5	Fresh ground beef	4.3
2% milk	4.1	Veal	2.7
Butter fat	6.1	Lamb	5.8
Condensed milk	7.0	Pork	0.6
Cultured buttermilk	5.4	Chicken	0.9
Butter	4.7	Fresh ground turkey	2.6
Sour cream	4.6	Salmon	0.3
Ice cream	3.6	Egg yolk	0.6
Low-fat yoghurt	4.4		
Custard style yoghurt	4.8	Vegetable oils	
Plain yoghurt	4.8	Safflower oil	0.7
Frozen yoghurt	2.8	Sunflower oil	0.4
Medium Cheddar	4.1		
American processed	5.0		

Adapted from MacDonald (2000).

3. Role of CLA in health and disease

The first health benefit of CLA was discovered rather accidentally in the 1980s when Pariza and Hargraves (1985) were investigating the carcinogenic properties of grilled beef. To their surprise and contrary to their expectations, a fatty acid present in grilled beef exhibited anticarcinogenic rather than procarcinogenic properties. Since that discovery substantial data, mainly from cell line and animal studies, indicate that individual isomers of CLA, mainly *c*9, *t*11 and *t*10, *c*12, have numerous health benefits including reduction of carcinogenesis, atherosclerosis, inflammation, obesity and diabetes, as well as increased growth and bone formation (Kritchevsky *et al.* 2000; Bassaganya-Riera *et al.* 2002; Whigham *et al.* 2002; Miller *et al.* 2003; Pariza *et al.* 2003; Terpstra 2004). Comparatively, there is very limited literature on the effects of CLA on human health, and the dramatic beneficial effects seen in some animal models have not yet been reflected in human studies. Lack of consistent data in human studies could partly be due to the fact that human trials are different in terms of the type of controls used, dietary monitoring, metabolic state of the subjects, dosage and period of CLA supplementation. The reported effects of CLA include the following (Cook *et al.* 1993; Miller *et al.* 1994; Pariza *et al.* 2001; Whigham *et al.* 2001; Bassaganya-Riera *et al.* 2002; Noone *et al.* 2002; Whigham *et al.* 2002; Pariza *et al.* 2003; Albers *et al.* 2003; Bassaganya-Riera *et al.* 2003; Belury 2003; Henrikssen *et al.* 2003; Kritchevsky 2003; Nagao *et al.* 2003):

- Anticarcinogenic
- Enhances immune function
- Reduces inflammation
- Reduces catabolic effects of immune stimulation
- Reduces asthma in animal models
- Reduces atherosclerosis
- Enhances growth of young rodents
- Reduces body fat gain
- Enhances lean body mass gain
- Reduces negative effects of weight-loss diets
- Reduces symptoms of diabetes in some models
- Reduces hypertension.

It is clear from recent studies that the two main isomers can have different effects on metabolism and cell function and that they act through different cell signalling pathways (Park *et al.* 2000; Yang and Cook 2003; Valeille *et al.* 2004; Storkson *et al.* 2005). However, two main mechanistic theories have been proposed to explain how CLA is believed to exert its physiological effects. One theory suggests that CLA reduces the amount of arachidonic acid (AA) in phospholipids, resulting in a reduced amount of AA released from cell membranes and thereby reduced production of downstream eicosanoid products such as prostaglanding E_2 (PGE_2) and leukotriene B_4 (LTB_4), which are involved in inflammation and cancer (Fulton 1998; Funk 2001). The second model involves CLA as a regulator of the expression of genes known to control, for example, such processes as lipid metabolism, energy balance, apoptosis and immune function (Pariza *et al.* 2000; Belury 2002).

The beneficial effects of CLA on cancer are probably the most studied of all health benefits associated with the fatty acid. Feeding a mixture of CLA isomers has been shown to inhibit chemically-induced tumours of the mammary gland, skin, colon, and forestomach in several animal models and human cell culture studies (for reviews see Bhattacharya *et al.* 2006; Kelley *et al.* 2007). The mechanisms for this anticancer activity are proposed to include an alteration of lipid peroxidation, tissue fatty acid composition, eicosanoid metabolism, gene expression, cell cycle regulation, cell proliferation and apoptosis (Kelley *et al.* 2007). The inhibition of tumorigenesis has been shown to be dependent upon the dietary concentrations of the CLA mixtures, varying from 0.05 to 1% (wt/wt; Kelley *et al.* 2007). Interestingly, in humans these dietary concentrations would be obtainable either through supplementation or via a significant change in the intake of CLA-rich foods.

It has also been reported that the *c*9, *t*11 and *t*10, *c*12 CLA isomers exert distinct effects on immune function. Proinflammatory cytokines (TNF-α, IL-6, IL-1 etc.), anti-inflammatory cytokines (IL-4, IL-10 etc.), eicosanoids (prostaglandins, leukotrienes) and nitric oxide (NO) are key inflammatory mediators that are regulated by the dietary intake of polyunsaturated fatty acids (PUFA), including n-6 and n-3 fatty acids. CLA has been shown to inhibit the production of proinflammatory cytokines and conversely to stimulate the production of anti-inflammatory cytokines in different animal models (Akahoshi *et al.* 2002; Yang and Cook 2003; Changhua *et al.* 2005). TNF-α is a key regulatory cytokine in inflammation; it promotes connective tissue degradation and is involved in the induction of a chronic inflammatory state (Toussirot and Wendling 2004). Therefore, inhibition of TNF-α is of crucial importance in chronic inflammatory conditions such as inflammatory bowel disease and rheumatoid arthritis.

There is a growing body of evidence that CLA can decrease the extent of atherosclerosis in experimental animals. Dietary CLA supplementation of rabbits (Kritchevsky *et al.* 2000; Kritchevsky *et al.* 2002) and hamsters (Nicolosi *et al.* 1997; Wilson *et al.* 2000) have been shown to decrease the development of early atherosclerotic lesions. When CLA (0.5g/day) was added to a hypercholesterolaemic diet and fed to rabbits for 12 weeks, serum triglycerides and low-density lipoprotein cholesterol levels were significantly reduced compared with rabbits fed a diet with no CLA supplemenation. Importantly for heart disease risk, aortas of rabbits fed the CLA-containing diet exhibited less atherosclerotic plaque formation (Lee *et al.* 1994). As little as 0.05% CLA has been shown to reduce the severity of atherosclerotic lesions in rabbits (Kritchevsky *et al.* 2002).

CLA has also attracted interest in the scientific community because of its promising effects on body composition including the reduction of body fat mass and increases in lean mass. Because of the increasing prevalence and incidence of obesity, and given that obesity is associated with a high risk of developing diabetes and cardiovascular disease, management of body weight is considered an important approach to maintenance of good health. Dietary CLA has been shown to reduce body fat effectively and consistently in mice, with a reduction of up to 70% obtained compared to controls (DeLany *et al.* 1999; Pariza *et al.* 2003). However, CLA has been shown to be less effective in rats, pigs, and in humans (Dugan *et al.* 2004; Mirand *et al.* 2004; Terpstra 2004; Wang and Jones

2004). Relatively few human trials have been conducted to determine the effects of CLA on body mass and composition. The results of these studies have generally shown a reduction in body fat, but not body mass, with CLA supplementation (Blankson *et al.* 2000; Smedman and Vessby 2001; Thom *et al.* 2001). Differences in effects of CLA on body fat reduction in humans compared to animals may be due to the relatively low doses used in human studies compared to animal trials. For example, mice are generally fed approximately five times more CLA than humans (per kilogram body weight). Moreover, most animal studies have been performed in young growing mice or rats, whereas studies in humans are mostly in mature volunteers.

4. Enrichment of CLA by dietary and non dietary manipulation

Methods by which the CLA content of milk fat and oils could be enhanced are now the focus of many studies since CLA enrichment could be very advantageous in light of the many scientifically-proven beneficial effects of this unique fatty acid. Altered feeding conditions and ruminal fermentation are strategies used for increasing the CLA content of dairy products (Chouinard *et al.* 2001; Duckett *et al.* 2002). Dietary manipulation has been recognised for decades as being the most successful strategy for elevating the CLA content of milk fat (Booth *et al.* 1935), particularly supplementation of the ruminant diet with plant oils, fish oils, animal fats, synthetic CLA supplements and forage. Supplementing the ruminant diet with polyunsaturated oils that contain either linoleic acid (e.g. corn oil and sunflower oil) or linolenic acid (e.g. linseed oil) have been shown to increase the CLA content of milk fat substantially (Dhiman *et al.* 1999; Griinari and Bauman 1999; Bauman *et al.* 2000; Jones *et al.* 2000). Milk from cows fed sunflower oil (53 g/kg dry weight) for two weeks contained a higher CLA content (24.4 mg/g fat) than milk from cows fed similar amounts of linseed oil (16.7 mg/g fat) or peanut oil (13.3 mg/g fat). Sunflower oil contains greater amounts of linoleic acid (69.4 g/100 g) than peanut (30.2 g/100g) or linseed oil (15.4 g/100g); therefore, increasing the content of this precursor fatty acid in the diet led to the highest concentration of CLA in milk (Kelly *et al.* 1998). Linseed oil is rich in linolenic acid, another substrate for CLA formation in milk. Dietary linolenic acid also results in increased concentration of vaccenic acid in the rumen, which is then converted to CLA via the Δ^9-desaturase enzyme in the mammary gland (Griinari *et al.* 2000; Kay *et al.* 2004). Several studies have also shown that pasture feeding significantly and rapidly increases milk fat CLA concentrations in lactating dairy cows (Stanton *et al.* 1997; Lawless *et al.* 1998; Dhiman *et al.* 1999). Cows receiving all of their daily feed as pasture produced milk fat with a greater CLA content (22 mg/g fat) than cows receiving only one-third (8 mg/g fat) or two-thirds (14 mg/g fat) of their daily feed as pasture (Dhiman *et al.* 1999). Moreover, dietary supplementation of lactating ruminants with fish oils has been shown to result in an elevation in milk fat CLA (Rego *et al.* 2005; Allred *et al.* 2006). Fish oils are rich in long chain PUFA, such as eicosapentaenoic acid (EPA) and docosahexaenoic acid (DHA). EPA and DHA inhibit reductase enzymes that catalyse the conversion of vaccenic acid to stearic acid in the rumen (Chow *et al.* 2004; Larsson *et al.* 2005). This leads to an accumulation of vaccenic acid in the rumen, which is then transformed to *c*9, *t*11 CLA in the mammary gland by the action of Δ^9-desaturase (Griinari and Bauman 1999; Chow *et al.* 2004).

Microbial synthesis of CLA by food-grade strains also provide a means of enriching the CLA content of food products. Lactococci, streptococci, enterococci and propionibacteria are important lactic acid bacteria (LAB) that are commonly used as dairy starter cultures and that play a critical role in the manufacture of fermented dairy products such as yoghurt, buttermilk, lactic butter and cheese. A number of studies have shown that these bacteria possess the ability to produce CLA from free linoleic acid (Jiang *et al.* 1998; Lin *et al.* 1999; Alonso *et al.* 2002), which has opened up new opportunities for the development of fermented food products enriched in CLA. Jiang *et al.* (1998) investigated the production of CLA by dairy starter cultures *in vitro* using 19 different strains of lactobacilli, lactococci, streptococci and propionibacteria. Two strains of *Propionibacterium freudenreichii* ssp. *freudenreichii* and one strain of *P. freudenreichii* ssp. *sheramnii* were capable of converting free linoleic acid to extracellular CLA. The highest level of CLA formed in the media was 0.27 mg/mL by the strain *P. freudenreichii* ssp. *freudenreichii*. This represented a two- to three-fold increase in CLA formation compared with other strains examined; of the different isomers, *c*9, *t*11 represented more than 70% of the total CLA formed. Kim and Liu (2002) screened 14 LAB for CLA production using sunflower oil as a substrate. Among the strains examined, *Lactococcus lactis* I-01 exhibited the highest CLA-producing ability. The CLA content attained through lactic acid fermentation using this strain was 2.3 times higher (11 mg/g fat) than the CLA content of normal milk fat.

5. CLA production by human intestinal bacteria

Evidence of CLA formation by human intestinal bacteria was first obtained indirectly by Chin *et al.* (1994), who observed that germ-free rats had a lower incorporation of CLA into different tissues than conventional rats. This led to speculation that intestinal bacteria could produce CLA. Consequently, CLA status may be increased not only by consuming CLA-enriched dairy products, but also by modifying the gut microbiota through the introduction of high CLA-producing bacteria. To date, a number of non-ruminant cultures are known to be capable of microbial *c*9, *t*11 CLA synthesis, including strains that are natural residents of the human GIT such as *Lactobacillus*, *Bifidobacterium*, *Propionibacterium*, *Pediococcus*, *Enterococcus*, and *Lactococcus* (Jiang *et al.* 1998; Ogawa *et al.* 2001; Alonso *et al.* 2002; Kishino *et al.* 2002; Coakley *et al.* 2003; Oh *et al.* 2003; Rosberg-Cody *et al.* 2004; Ewaschuk *et al.* 2006; Lee *et al.* 2006; Barrett *et al.* 2007; Lee *et al.* 2007). These bacteria convert free linoleic acid to *c*9, *t*11 CLA in a similar manner to ruminant bacteria via the action of the enzyme linoleic acid isomerase (Lin *et al.* 2002). In contrast to *c*9, *t*11 CLA, *t*10, *c*12 CLA-producing bacteria have only been reported in a few studies (Alonso *et al.* 2002; Ewaschuk *et al.* 2006; Lee *et al.* 2006; Lee *et al.* 2007). Species of lactobacilli and bifidobacteria of human origin that have been reported to produce CLA are shown in Table 2 and some of these are discussed below.

Table 2. Lactobacilli and bifidobacteria species of human origin that have been reported to produce CLA

Species	Reference
B. breve	Oh *et al.* 2003; Coakley *et al.* 2003; Rosberg-Cody *et al.* 2004; Ewaschuk *et al.* 2006; Barrett *et al.* 2007
B. psuedocatenulatum	Oh *et al.* 2003; Coakley *et al.* 2003; Barrett *et al.* 2007
B. bifidum	Coakley *et al.* 2003; Rosberg-Cody *et al.* 2004
B. infantis	Coakley *et al.* 2003; Ewaschuk *et al.* 2006; Barrett *et al.* 2007
B. dentium	Coakley *et al.* 2003; Barrett *et al.* 2007
B. longum	Ewaschuk *et al.* 2006; Barrett *et al.* 2007
B. acodelescentis	Coakley *et al.* 2003
B. angulatum	Coakley *et al.* 2003
B. lactis	Coakley *et al.* 2003
L. acidophilus	Lin *et al.* 1999; Ogawa *et al.* 2001; Kim and Liu 2002; Kishino *et al.* 2002; Alonso *et al.* 2003; Lin *et al.* 2005
L. plantarum	Kishino *et al.* 2002; Kim and Liu 2002; Ewaschuk *et al.* 2006; Lee *et al.* 2007
L. rhamnosus	Kishino *et al.* 2002; Lee *et al.* 2006
L. casei	Alonso *et al.* 2003; Ewaschuk *et al.* 2006
L. paracasei	Kishino *et al.* 2002; Lee *et al.* 2006
L. pentosus	Kishino *et al.* 2002; Lee *et al.* 2006
L. delbrueckii ssp. *bulgaricus*	Lin *et al.* 1999; Lin *et al.* 2005
L. delbrueckii ssp. *bulgaricus*	Lin *et al.* 1999; Lin *et al.* 2005
L. delbrueckii ssp. *lactis*	Lin *et al.* 1999
L. brevis	Kishino *et al.* 2002

5.1 CLA production by lactobacilli in vitro

Several studies have reported that lactobacilli possess the ability to produce CLA when cultivated in media containing linoleic acid. For example, Kishino *et al.* (2002) assayed more than 250 strains from 14 genera of LAB for their CLA-producing abilities when cultivated in MRS (de Man, Rogosa, Sharpe) medium containing 0.6 mg/mL linoleic acid. Four *Lactobacillus* strains included in their study produced more than 1.4 mg total CLA/mL reaction mixture: *L. rhamnosus* AKU 1124, *L. rhamnosus* AKU 1137, *L. plantarum* AKU 1009a and *L. plantarum* JCM 1551, of which *L. plantarum* AKU 1009a produced as much as 3.41 mg/mL. After optimisation of reaction conditions, washed cells of *L. plantarum* AKU 1009a produced as much as 40 mg/mL CLA. The CLA produced was a mixture of two isomers, i.e. *c*9, *t*11 CLA and *t*9, *t*11 CLA (Kishino *et al.* 2002). Ogawa *et al.* (2001) used washed cells of *L. acidophilus* AKU 1137, which produced high amounts of CLA (4.9 mg/mL, a mixture of *c*9, *t*11 and *t*9, *t*11) when incubated at a linoleic acid concentration of 5 mg/mL. In the studies by Kishino *et al.* (2002) and Ogawa *et al.* (2001), almost all of the CLA produced was contained within the cells or associated with the cells as free fatty acids. Based on these observations, the use of washed cells would appear to be an effective method for the natural production of CLA on an industrial scale. Kim and Liu (2002) tested eight different strains of *L. acidophilus*, *L. bulgaricus*, *L. plantarum*, *L. helveticus* and *L. johnsonii* for CLA production in both MRS medium and whole milk at a linoleic acid concentration of 0.1 mg/mL. When grown in MRS medium only two strains (*L. acidophilus* 96 and *L. plantarum* 4191) produced CLA, while growth in whole milk resulted in additional CLA production by two other strains (*L. acidophilus* 56 and *L. acidophilus* ATCC 43121). Growth in whole milk also resulted in increased CLA production by *L. plantarum* 4191. Alonso *et al.* (2002) tested four different *Lactobacillus* strains of human intestinal origin, i.e. two strains each of *L. acidophilus* and *L. casei*, for their ability to produce CLA from free linoleic acid in MRS supplemented with different concentrations of linoleic acid (0.05, 0.1, 0.2 and 0.5 mg/mL). In contrast to the above studies by Kishino *et al.* (2002) and Ogawa *et al.* (2001), this group removed all of the cells and assayed only the free CLA in the spent growth media. All of the four cultures tested converted linoleic acid to CLA with *c*9, *t*11, *t*10, *c*12 and *t*9, *t*11 being formed, of which *c*9, *t*11 comprised more than 90% of the total CLA. Ewaschuk *et al.* (2006) assessed whether a mixture of probiotic strains VSL#3 (*L. casei*, *L. plantarum*, *L. acidophilus*, *L. delbrueckii* ssp. *bulgaricus*, *Bifidobacterium infantis*, *Bifidobacterium breve*, *Bifidobacterium longum* and

Streptococcus thermophilus) could convert linoleic acid to CLA after 24 h incubation in the presence of 0.5 mg/mL linoleic acid. As a result, all the bacterial strains within the probiotic mixture were found to convert linoleic acid to CLA to varying degrees. Interestingly, all strains produced c9, *t*11 and *t*10, *c*12 CLA. Lee *et al.* (2006) analysed the crude enzyme prepared from washed cells of three *Lactobacillus* strains (*L. rhamnosus* PL60, *L. paracasei* IFO 12004 and *L. pentosus* IFO 12011) for CLA production. They found that two of these strains, i.e. *L. rhamnosus* PL60 and *L. pentosus* IFO 12011, produced both *c*9, *t*11 and *t*10, *c*12 CLA, while *L. paracasei* IFO 12004 produced the *c*9, *t*11 isomer but not *t*10, *c*12 CLA. Moreover, the same group isolated *L. plantarum* PL62 from infant faeces, which produced c9, *t*11 and *t*10, *c*12 CLA (Lee *et al.* 2007). After incubation with 0.1% linoleic acid, the culture supernatant of PL62 contained 26.8 μg/mL *c*9, *t*11 CLA and 6.4 μg/mL *t*10, *c*12 CLA, whereas the amounts of each isomer in the reaction mixture of the crude enzyme of PL62 were 1252 μg *c*9, *t*11 CLA/mg protein and 1395 μg *t*10, *c*12 CLA/mg protein, respectively. Studies on the production of CLA by lactobacilli have generally found the majority of the CLA produced to be associated with the cells or located within these. Storage of CLA by the cells in this manner increases the potential for the use of lactobacilli as probiotic vectors for delivery of CLA to the human GIT.

5.2 CLA production by bifidobacteria

It was reported that *Bifidobacterium* species of human origin can conjugate linoleic acid and produce CLA (Coakley *et al.* 2003). In this study, a number of *Bifidobacterium* strains isolated from adult and infant intestine were assessed for their ability to produce CLA. After incubation for 48 h in MRS medium with 0.55 mg/mL linoleic acid it was found that *B. breve*, *B. dentium* and *B. lactis* formed relatively high amounts of *c*9, *t*11 CLA (0.12–0.36 mg/mL), whereas the other bifidobacteria tested (*B. adolescentis*, *B. bifidum* and *B. infantis*) did not produce any CLA. The best producer in this study was *B. breve* NCIMB 702258 (Figure 2). A number of other studies have also investigated the production of CLA by bifidobacteria of human origin. Oh *et al.* (2003) screened faecal samples derived from Korean breast-fed infants for CLA-producing isolates. They found two strains with efficient CLA-producing capabilities, i.e. *B. breve* KCTC 10462 and *B. pseudocatenulatum* KCTC 10208. In both cases,

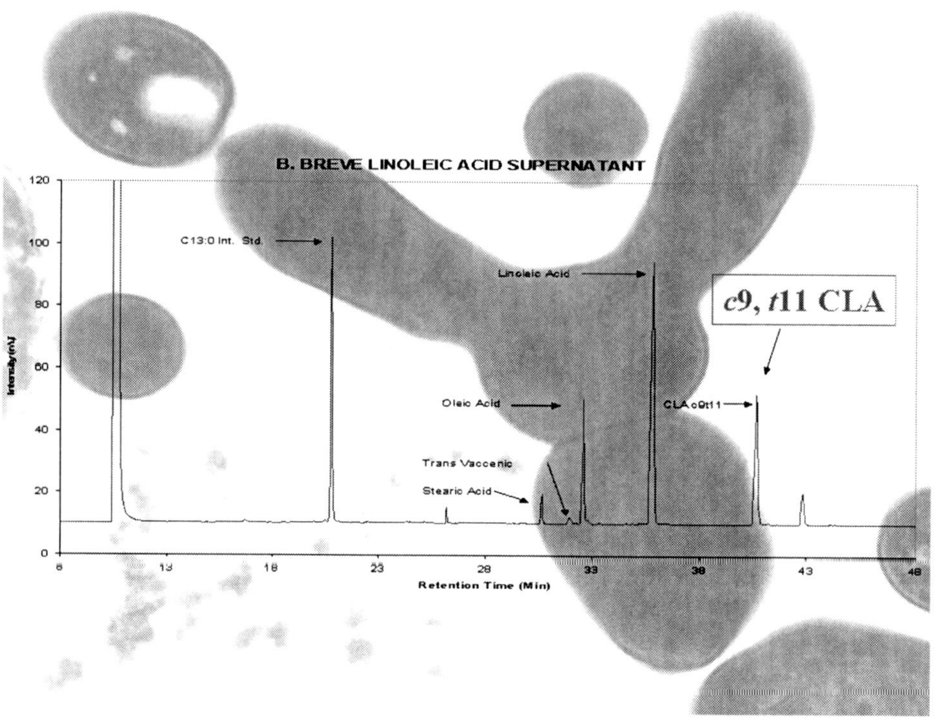

Figure 2. Chromatogram showing *c*9, *t*11 conjugated linoleic acid (CLA) production by *Bifidobacterium breve* NCIMB 702258. After incubation for 48 h in MRS medium with 0.55 mg/mL of linoleic acid, *B. breve* NCIMB 702258 produced relatively high amounts of *c*9, *t*11 CLA (approx. 0.36 mg/mL; Coakley *et al.* 2003). An electron micrograph of *B. breve* NCIMB 702258 is shown in the background.

the major CLA isomer produced was the *c*9, *t*11 CLA. The quantity of CLA in the supernatant produced by *B. breve* and *B. pseudocatenulatum* from 0.50 mg/mL linoleic acid was 0.16 and 0.14 mg/mL, respectively. Rosberg-Cody *et al.* (2004) screened 46 *Bifidobacterium* isolates from neonatal faecal material for CLA production. The most efficient producers belonged to the species *B. breve*, of which two different strains converted 29 and 27% of the free linoleic acid to the *c*9, *t*11 isomer. In addition, a strain of *B. bifidum* was found to have a conversion rate of 18%. Barrett *et al.* (2007) screened 870 isolates from 58 faecal samples obtained from a diverse human population for CLA production. Of these, 18 distinct CLA-producing *Bifidobacterium* strains were identified from 15 of the 58 samples. Nine of these CLA-producing strains belonged to *B. longum*, six to *B. breve*, and a single strain belonged to each of *B. infantis*, *B. dentium* and *B. catenulatum*. In all cases, *c*9, *t*11 CLA was the predominant isomer in the culture supernatant; the *t*9, *t*11 CLA isomer was also generated, but at much lower levels. Nine of the strains converted more than 20% of linoleic acid to CLA, and one of the strains (*B. breve*) converted as much as 75%. Five of the CLA-producing strains that were able to convert more than 20% of linoleic acid to CLA belonged to the species *B. breve*, and the remaining four were identified as *B. longum* strains. The source of the CLA-producing bifidobacteria was a further point of interest in this study. In most previous studies workers had isolated CLA-producing bifidobacteria from the faecal material of neonates (Coakley *et al.* 2003; Oh *et al.* 2003; Rosberg-Cody *et al.* 2004), which led to the assumption that populations of bifidobacteria with the ability to produce CLA develop shortly after birth and therefore may play an important role in the health of the neonate. The best source of CLA-producing bifidobacteria in the study by Barrett *et al.* (1997), in terms of both conversion of linoleic acid to CLA and prevalence of CLA-producing strains within a group, was the faecal material of elderly patients infected with the bacterium *Clostridium difficile*. Only 18% of the healthy infants screened for CLA-producing bifidobacteria harboured CLA-producing strains, while 35% of the *C. difficile* patients screened harboured CLA-producing strains (Barrett *et al.* 2007).

5.3 Biosynthesis of CLA by human bacteria in vivo

To date, few studies have investigated the *in vivo* production of CLA using bacteria of human origin. Kamlage *et al.* (1999) concluded that intestinal bacteria do not supply rats with systemic CLA. They reported that CLA was produced from linoleic acid when human faeces were incubated *in vitro*, but that no CLA was produced when the same faecal bacteria were introduced into germ-free rats. They reported that the difference between the *in vivo* and *in vitro* results was due to the fact that CLA production was inhibited by glucose (Kamlage *et al.* 2000). Nevertheless, two recent studies have reported the *in vivo* production of CLA using two strains of human origin, i.e. *Lactobacillus rhamnosus* PL60 and *L. plantarum* PL62 (Lee *et al.* 2006, 2007). Both of these studies demonstrated the *in vivo* production of *t*10, *c*12 CLA by these strains in obese mice. Interestingly, when five mice received 1 × 10^9 colony forming units (CFU) of *L. rhamnosus* PL60 daily for 3 weeks, 1.9 µg/mL of *t*10, *c*12 CLA was detected in the sera, while this isomer was not detected in the control group (Lee *et al.* 2006). This finding opens up possibilities to provide CLA directly to the intestinal lumen. Given that CLA has been shown to exhibit antiproliferative and anti-inflammatory effects on colonocytes (Kemp *et al.* 2003; Bassaganya-Riera *et al.* 2004), the provision of CLA directly to the intestinal lumen may be considered beneficial, particularly for inflammatory bowel disorders such as ulcerative colitis and Crohn's disease (Henning *et al.* 2001).

6. CLA production by gut microbiota – a novel probiotic trait?

The presence of *Bifidobacterium* and *Lactobacillus* species in the GIT has long been associated with a number of beneficial health effects, including reduction in the duration of antibiotic-associated diarrhoea, modulation of the mucosal immune system, alleviation of lactose intolerance and improvement of GIT disturbances (Rowland *et al.* 1998; Marteau *et al.* 2003; Brown and Valiere 2004; Hart *et al.* 2004; Chen and Walker; 2005; Jonkers *et al.* 2007). Since they exert such desirable health effects, bifidobacteria and lactobacilli are increasingly being used as probiotics, i.e. "beneficial bacteria, which when administered in adequate amounts confer a health benefit on the host" (FAO/WHO 2002), in functional food and pharmaceutical products.

The ability of bifidobacteria and lactobacilli to produce bioactive fatty acid metabolites such as CLA could be linked to their health-promoting properties. This has been reported in a few recent studies. For example, Coakley *et al.* (2006) investigated the antiproliferative activities of *c*9, *t*11 and *t*9, *t*11 CLA, produced by *B. breve* NCIMB 702258, using two different human colon cancer cell lines (SW480 and HT-29). Both of these microbially-produced isomers showed a strong antiproliferative effect against the cell lines examined, with *t*9, *t*11 having a more potent antiproliferative effect than the *c*9, *t*11 isomer. In addition, Rosberg-Cody *et al.* (2007) cloned and overexpressed linoleic acid isomerase from *Propionibacterium acnes* into *Lactococcus lactis* and *Escherichia coli* and assessed whether the bacterially produced *t*10, *c*12 isomer had an antiproliferative effect on SW480 colon cancer cells. The fatty acids generated from both *L. lactis* and *E. coli* were found to contain a mixture of linoleic acid and *t*10, *c*12 CLA at a ratio of approximately 1.35:1. Following five days of incubation of SW480 cells with 5–20 μg/mL of *t*10, *c*12 CLA, there was a significant reduction in cell growth compared with the linoleic acid control. After treatment with the highest concentration (20 μg/mL) of the *t*10, *c*12 isomer, cell viability was reduced to 7.9% (*L. lactis* CLA) and 19.6% (*E. coli* CLA), compared with 95.4% (control LA) and 31.7% (pure *t*10, *c*12 CLA). In a separate study, the anticarcinogenic effect of CLA (*c*9, *t*11 and *t*10, *c*12) produced by the probiotic mixture VSL#3 on HT-29 and Caco-2 cell lines was investigated (Ewaschuk *et al.* 2006). Conditioned medium containing the probiotic-produced CLA reduced viability and induced apoptosis in both cell lines. In addition, incubation of murine faeces with linoleic acid after administration of VSL#3 yielded 100-fold more CLA compared to faeces collected prior to VSL#3 feeding.

A recent study by Lee *et al.* (2006) reported that *t*10, *c*12 CLA-producing *L. rhamnosus* PL60 increased the *t*10, *c*12 CLA concentration in sera and consequently reduced body weight without reducing energy intake, and caused a significant, specific reduction in white adipose tissue in diet-induced obese mice. They also demonstrated that another *t*10, *c*12 CLA-producing strain of human origin—*L. plantarum* PL62—could exert similar antiobesity activity in mice, reducing the weight of epididymal, inguinal, mesenteric and perirenal white adipose tissue and significantly reducing both the blood levels of total glucose and body weight (Lee *et al.* 2007). Results from these studies indicate that bacterially-produced CLA would likely have a positive effect on human health. Therefore, it is tempting to suggest that the ability of some strains of *Bifidobacterium* and *Lactobacillus* to produce these bioactive metabolites may be associated with the beneficial effects of these bacteria within the human GIT.

7. Conclusions

Research to date has demonstrated that microbially-produced CLA has the potential to enhance host health and even alleviate disease. Therefore, generation of functional foods containing this bioactive lipid, and/or the presence of bacteria that produce CLA in the human GIT could increase CLA status and may therefore play an important role in maintaining human health. Introducing probiotics with CLA-producing capability instead of using a pure form of CLA would involve fewer safety concerns and would exert additional beneficial effects, such as promoting intestinal health and the immunoenhancing activity of probiotics in general. Furthermore, compared with dietary CLA, which is primarily absorbed in the small intestine, a probiotic may continuously provide CLA to the large intestine where it would exert local beneficial effects on colonocytes.

8. Acknowledgements

This work was funded by SFI funds, and in part by EU Project QLK1-2002-02362 and the Irish Government under the National Development Plan 2000–2006. Rebecca Wall is an APC student funded by the Alimentary Pharmabiotic Centre (APC).

9. References

Akahoshi, A., Goto, Y., Murao, K. *et al.* 2002. Conjugated linoleic acid reduces body fat and cytokine levels of mice. *Bioscience, Biotechnology, and Biochemistry* **66**(4): 916-920.

Albers, R., van der Wielen, R.P.J., Brink, E.J., Hendriks, H.F.J., Dorovska-Taran, V.N. and Mohede, I.C.M. 2003. Effects of *cis*-9, *trans*-11 and *trans*-10, *cis*-12 conjugated linoleic acid (CLA) isomers on immune function in healthy men. *European Journal of Clinical Nutrition* **57**: 595-603.

Allred, S.L., Dhiman, T.R., Brennand, C.P., Khanal, R.C., McMahon, D.J. and Luchini, N.D. 2006. Milk and cheese from cows fed calcium salts of palm and fish oil alone or in combination with soybean products. *Journal of Dairy Science* **89**: 234-248.

Alonso, L., Cuesta, E.P. and Gilliand, S.E. 2002. Production of free conjugated linoleic acid by *Lactobacillus acidophilus* and *Lactobacillus casei* of human intestinal origin. *Journal of Dairy Science* **86**: 1941-1946.

Barrett, E., Ross, R.P., Fitzgerald, G.F. and Stanton, C. 2007. Rapid screening method for analyzing the conjugated linoleic acid production capabilities of bacterial cultures. *Applied Environmental Microbiology* **73**: 2333-2337.

Bassaganya-Riera, J., Hontecillas, R. and Beitz, D.C. 2002. Colonic anti-inflammatory mechanisms of conjugated linoleic acid. *Clinical Nutrition* **21**: 451-459.

Bassaganya-Riera, J., Pogranichniy, R.M., Jobgen, S.C. *et al.* 2003. Conjugated linoleic acid ameliorates viral infectivity in a pig model of virally induced immunosuppression. *Journal of Nutrition* **133**: 3204-3214.

Bassaganya-Riera, J., Reynolds, K. and Martino-Catt, S. *et al.* 2004. Activation of PPARγ and δ by conjugated linoleic acid mediates protection from experimental inflammatory bowel disease. *Gastroenterology* **127**: 777-791.

Bauman, D.E., Barbano, D.M., Dwyer, D.A. and Griinari, J.M. 2000. Technical note: production of butter with enhanced conjugated linoleic acid for use in biomedical studies with animal models. *Journal of Dairy Science* **83**: 2422-2425.

Belury, M.A. 2002. Dietary conjugated linoleic acid in health: physiological effects and mechanisms of action. *Annual Review of Nutrition* **22**: 505-531.

Belury, M.A. 2003. Conjugated linoleic acids in type 2 diabetes mellitus: implications and potential mechanisms. In: Sebedio, J.L., Christie, W.W. and Adlof, R., editors. *Advances in conjugated linoleic acid research*, volume 2: 302-315. AOCS Press, Champaign, Illinois, USA.

Bhattacharya, A., Banu, J., Rahman, M., Causey, J. and Fernandes, G. 2006. Biological effects of conjugated linoleic acids in health and disease. *Journal of Nutritional Biochemistry* **17**: 789-810.

Blankson, H.J., Stakkestad, A., Fagertun, H., Thom, E., Wadstein, J. and Gudmundsen, O. 2000. Conjugated linoleic acid reduces body fat mass in overweight and obese humans. *Journal of Nutrition* **130**: 2943-2948.

Booth, R.G., Kon, S.K., Dann, W.J. and Moore, T. 1935. A study of seasonal variation in butter fat: a seasonal spectroscopic variation in the fatty acid fraction. *Biochemistry Journal* **29**: 133-137.

Brown, A.C. and Valiere, A. 2004. Probiotics and medical nutrition therapy. *Nutrition in Clinical Care* **7**: 56-68.

Changhua, L., Jindong, Y., Defa, L., Lidan, Z., Shiyan, Q. and Jiajun, X. 2005. Conjugated linoleic acid attenuates the production and gene expression of proinflammatory cytokines in weaned pigs challenged with lipopolysaccharide. *Journal of Nutrition* **135**: 239-244.

Chen, C.C. and Walker, W.A. 2005. Probiotics and prebiotics: role in clinical disease states. *Advances in Pediatrics* **52**: 77-113.

Chin, S. F., Storkson, J.M., Liu, W., Albright, K.J. and Pariza, M.W. 1994. Conjugated linoleic acid (9,11- and 10,12-octadecadienoic acid) is produced in conventional but not germ free rats fed linoleic acid. *Journal of Nutrition* **124**: 694-701.

Chouinard, P.Y., Corneau, L., Butler, W.R., Chillard, Y., Drackley, J.K. and Barman, D.E. 2001. Effect of dietary lipid source on conjugated linoleic acid concentrations in milk fat. *Journal of Dairy Science* **84**: 680-690.

Chow, T.T., Fievez, V., Moloney, A.P., Raes, K., Demeyer, D. and De Smet, S. 2004. Effect of fish oil on *in vitro* rumen lipolysis, apparent biohydrogenation of linoleic and linolenic acid and accumulation of biohydrogenation intermediates. *Animal Feed Science and Technology* **117**: 1-12.

Coakley, M., Ross, R.P., Nordgren, M., Fitzgerald, G., Devery, R. and Stanton, C. 2003. Conjugated linoleic acid biosynthesis by human-derived *Bifidobacterium* species. *Journal of Applied Microbiology* **94**: 138-145.

Coakley, M., Johnson, M.C., McGrath, E. *et al.* 2006. Intestinal bifidobacteria that produce *trans*-9, *trans*-11 conjugated linoleic acid: a fatty acid with antiproliferative activity against human colon SW480 and HT-29 cancer cells. *Nutrition and Cancer* **56**(1): 95-102.

Cook, M.E., Miller, C.C., Park, Y. and Pariza, M. 1993. Immune modulation by altered nutrient metabolism: nutritional control of immune-induced growth depression. *Poultry Science* **72**: 1301-1305.

Coxam, V. 2005. Inulin-type fructans and bone health: state of the art and perspectives in the management of osteoporosis. *British Journal of Nutrition* **93**: S111-S123.

DeLany, J.P., Blohm, F., Truett, A.A., Scimeca, J.A. and West, D.B. 1999. Conjugated linoleic acid rapidly reduces body fat content in mice without affecting energy intake. *American Journal of Physiology* **276**: R1172-R1179.

Dhiman, T.R., Anand, G.R., Satter, L.D. and Pariza, M.W. 1999. Conjugated linoleic acid content of milk from cows fed different diets. *Journal of Dairy Science* **82**: 2146-2156.

Duckett, S.K., Andrae, J.G. and Owens, F.N. 2002. Effect of high-oil corn or added corn oil on ruminal biohydrogenation of fatty acids and conjugated linoleic acid formation in beef steers fed finishing diets. *Journal of Animal Science* **80**: 3353-3360.

Dugan, M.E., Aalhus, J.L. and Kramer, J.K. 2004. Conjugated linoleic acid pork research. *American Journal of Clinical Nutrition* **79**: 1212S-1216S.

Ewaschuk, J.B., Walter, J.W., Diaz, H. and Madsen, K.L. 2006. Bioproduction of conjugated linoleic acid by probiotic bacteria occurs *in vitro* and *in vivo* in mice. *Journal of Nutrition* **136**: 1483-1487.

Joint FAO/WHO Working group report on drafting guidelines for the evaluation of probiotics in food 2002. London, Ontario, Canada.

French, P., Stanton, C., Lawless, F. *et al.* 2000. Fatty acid composition, including conjugated linoleic acid, of intramuscular fat from steers offered grazed grass, grass silage, or concentrate-based diets. *Journal of Animal Science* **78**: 2849-2855.

Fulton, A.M. 1998. The role of eicosanoids in tumor metastasis. *Prostaglandins Leukotrienes and Essential Fatty Acids – Reviews* **34**: 229-237.

Funk, C.D. 2001. Prostaglandins and leukotrienes: advances in eicosanoid biology. *Science* **294**(5548): 1871-1875.

Griinari, J.M. and Bauman, D.E. 1999. Biosynthesis of conjugated linoleic acid and its incorporation into meat and milk in ruminants. In: Yurawecz, M.P., Mossoba, M.M., Kramer, J.K.G., Pariza, M.W. and Nelson, G.J., editors. *Advances in conjugated linoleic acid research,* volume 1: 180-200. AOCS Press, Champaign, Illinois, USA.

Griinari, J.M., Corl, B.A., Lacy, S.H., Chouinard, P.Y., Nurmela, K.V. and Bauman, D.E. 2000. Conjugated linoleic acid is synthesized endogenously in lactating dairy cows by delta (9)-desaturase. *Journal of Nutrition* **130**: 2285-2291.

Hart, A.L., Lammers, K., Brigidi, P. *et al.* 2004. Modulation of human dendritic cell phenotype and function by probiotic bacteria. *Gut* **53**(11): 1602-1619.

Hennig, B., Toborek, M. and McClain, C.J. 2001. High-energy diets, fatty acids and endothelial cell function: implications for atherosclerosis. *Journal of the American College of Nutrition* **20**: 97-105.

Henriksen, E.J., Teachey, M.K., Taylor, Z.C. *et al.* 2003. Isomer-specific actions of conjugated linoleic acid on muscle glucose transport in the obese Zucker rat. *American Journal of Physiology* **285**: E98-E105.

Hinnebusch, B.F., Meng, S., Wu, J.T., Archer, S.Y. and Hodin, R.A. 2002. The effects of short-chain fatty acids on human colon cancer cell phenotype are associated with histone hyperactivation. *Journal of Nutrition* **132**: 1012-1017.

Jiang, J., Björck, L. and Fonden, R. 1998. Production of conjugated linoleic acid by dairy starter cultures. *Journal of Applied Microbiology* **85**: 95-102.

Jiang, J., Wolk, A. and Vessby, B. 1999. Relation between the intake of milk fat and the occurrence of conjugated linoleic acid in human adipose tissue. *American Journal of Clinical Nutrition* **70**: 21-27.

Jones, D.F., Weiss, W.P. and Palmquist, D.L. 2000. Short communication: influence of dietary tallow and fish oil on milk fat composition. *Journal of Dairy Science* **83**: 2024-2026.

Jonkers, D. and Stockbrugger, R. 2007. Probiotics in gastrointestinal and liver diseases. *Alimentary Pharmacology and Therapeutics* **26**(2): 133-148.

Kamlage, B., Hartmann, L., Gruhl, B. and Blaut, M. 1999. Intestinal microorganisms do not supply associated gnotobiotic rats with conjugated linoleic acid. *Journal of Nutrition* **129**: 2212-2217.

Kamlage, B., Hartmann, L., Gruhl, B. and Blaut, M. 2000. Linoleic acid conjugation by human intestinal microorganisms is inhibited by glucose and other substrates *in vitro* and in gnotobiotic rats. *Journal of Nutrition* **130**: 2036-2039.

Kay, J.K., Mackle, T.R., Auldist, M.J., Thomson, N.A. and Bauman, D.E. 2004. Endogenous synthesis of *cis*-9, *trans*-11 conjugated linoleic acid in dairy cows fed pasture. *Journal of Dairy Science* **87**: 369-378.

Kelley, N.S., Hubbard, N.E. and Erickson, K.L. 2007. Conjugated linoleic acid isomers and cancer. *Journal of Nutrition* **137**(12): 2599-2607.

Kelly, M.L., Berry, J.R., Dwyer, D.A. *et al.* 1998. Dietary fatty acid sources affect conjugated linoleic acid concentrations in milk from lactating dairy cows. *Journal of Nutrition* **128**: 881-885.

Kemp, M.Q., Jeffy, B.D. and Romagnolo, D.F. 2003. Conjugated linoleic acid inhibits cell proliferation through a p53-dependent mechanism: effects on the expression of G1-restriction points in breast and colon cancer cells. *Journal of Nutrition* **133**: 3670-3677.

Kepler, C.R., Hirons, K.P., McNeill, J.J. and Tove, S.B. 1966. Intermediates and products of the biohydrogenation of linoleic acid by *Butyrivibrio fibrisolvens*. *Journal of Biological Chemistry* **241**: 1350-1354.

Kim, Y.J. and Liu, R.H. 2002. Increase of conjugated linoleic acid content in milk by fermentation with lactic acid bacteria. *Journal of Food Science* **67**(5): 1731-1737.

Kishino, S., Ogawa, J., Omura, Y., Matsumara, K. and Shimizu, S. 2002. Conjugated linoleic acid production from linoleic acid by lactic acid bacteria. *Journal of the American Oil Chemists' Society* **79**: 159-163.

Kritchevsky, D. 2003. Conjugated linoleic acid in experimental atherosclerosis. In: Sebedio, J.L., Christie, W.W. and Adlof, R., editors. *Advances in conjugated linoleic acid research*, volume 2: 293-301. AOCS Press, Champaign, Illinois, USA.

Kritchevsky, D., Tepper, S.A., Wright, S., Tso, P. and Czarnecki, S.K. 2000. Influence of conjugated linoleic acid (CLA) on establishment and progression of atherosclerosis in rabbits. *Journal of the American College of Nutrition* **19**: 472S-477S.

Kritchevsky, D., Tepper, S.A., Wright, S. and Czarnecki, S.K. 2002. Influence of graded levels of conjugated linoleic acid (CLA) on experimental atherosclerosis in rabbits. *Nutrition Research* **22**: 1275-1279.

Larsson, S.C., Bergkvist, L. and Wolk, A. 2005. High-fat dairy food and conjugated linoleic acid intakes in relation to colorectal cancer incidence in the Swedish Mammography Cohort. *American Journal of Clinical Nutrition* **82**(4): 894-900.

Lawless, F., Murphy, J.J., Harrington, D., Devery, R. and Stanton, C. 1998. Elevation of conjugated *cis*-9, *trans*-11-octadecadienoic acid in bovine milk because of dietary supplementation. *Journal of Dairy Science* **81**: 3259-3267.

Lee, K.N., Kritchevsky, D. and Pariza, M.W. 1994. Conjugated linoleic acid and atherosclerosis in rabbits. *Atherosclerosis* **108**: 19-25.

Lee, K., Paek, K., Lee, H.Y., Park, J.H. and Lee, Y. 2007. Antiobestity effect of *trans*-10, *cis*-12-conjugated linoleic acid-producing *Lactobacillus plantarum* PL62 on diet-induced obese mice. *Journal of Applied Microbiology* **103**: 1140-1146.

Lee, H.Y., Park, J.H., Seok, S.H. *et al.* 2006. Human originated bacteria, *Lactobacillus rhamnosus* PL60, produce conjugated linoleic acid and show anti-obesity effects in diet-induced obese mice. *Biochimica et Biophysica Acta* **1761**: 736-744.

Lin, T.Y., Lin, C.W. and Lee, C.H. 1999. Conjugated linoleic acid concentration as affected by lactic cultures and added linoleic acid. *Food Chemistry* **67**: 1-5.

Lin, T.Y., Lin, C.W. and Wang, Y.J. 2002. Linoleic acid isomerase activity in enzyme extracts from *Lactobacillus acidophilus* and *Propionibacterium freudenreichii* ssp. *shermanii*. *Journal of Food Science* **67**: 1502-1505.

Lin, T.Y., Hung, T.H. and Cheng, T.S.J. 2005. Conjugated linoleic acid production by immobilized cells of *Lactobacillus delbrueckii* ssp. *bulgaricus* and *Lactobacillus acidophilus*. *Food Chemistry* **92**: 23-28.

Lock, A.L. and Bauman, D.E. 2004. Modifying milk fat composition of dairy cows to enhance fatty acids beneficial to human health. *Lipids* **39**: 1197-1206.

MacDonald, H.B. 2000. Conjugated linoleic acid and disease prevention: a review of current knowledge. *Journal of the American College of Nutrition* **19**: 111S-118S.

Marteau, P. and Shanahan, F. 2003. Basic aspects and pharmacology of probiotics: an overview of pharmacokinetics, mechanisms of action and side-effects. *Best Practice Research Clinical Gastroenterology* **17**: 725-740.

Miller, C.C., Park, Y., Pariza, M.W. and Cook, M.E. 1994. Feeding conjugated linoleic acid to animals partially overcomes catabolic responses due to endotoxin injection. *Biochemical and Biophysical Research Communications* **198**: 1107-1112.

Miller, A., Stanton, C., Murphy, J. and Devery, R. 2003. Conjugated linoleic acid (CLA)-enriched milk fat inhibits growth and modulates CLA-responsive biomarkers in MCF-7 and SW480 human cancer cell lines. *British Journal of Nutrition* **90**: 877-885.

Mirand, P.P., Arnal-Bagnard, M.A., Mosoni, L., Faulconnier, Y., Chardigny, J.M. and Chilliard, Y. 2004. *Cis*-9, *trans*-11 and *trans*-10, *cis*-12 conjugated linoleic acid isomers do not modify body composition in adult sedentary or exercised rats. *Journal of Nutrition* **134**: 2263-2269.

Nagao, K., Inoue, N., Wang, Y.M. *et al.* 2003. The 10-*trans*, 12-*cis* isomer of conjugated linoleic acid suppresses the development of hypertension in Otsuka Long-Evans Tokushima fatty rats. *Biochemical and Biophysical Research Communications* **306**: 134-138.

Nicolosi, R.J., Rogers, E.J., Kritchevsky, D., Scimeca, J.A. and Huth, P.J. 1997. Dietary conjugated linoleic acid reduces plasma lipoproteins and early aortic atherosclerosis in hypercholesterolemic hamsters. *Artery* **22**: 266-277.

Noone, E.J., Roche, H.M., Nugent, A.P. and Gibney, M.J. 2002. The effect of dietary supplementation using isomeric blends of conjugated linoleic acid on lipid metabolism in healthy human subjects. *British Journal of Nutrition* **88**: 243-251.

Ogawa, J., Matsumura, K., Kishino, S., Omura, Y. and Shimizu, S. 2001. Conjugated linoleic acid accumulation via 10-hydroxy-12-octadecaenoic acid during microaerobic transformation of linoleic acid by *Lactobacillus acidophilus*. *Applied Environmental Microbiology* **67**(3): 1246-1252.

Oh, D.K., Hong, G.H., Lee, Y., Min, S., Sin, S. and Cho, S.K. 2003. Production of conjugated linoleic acid by isolated *Bifidobacterium* strains. *World Journal of Microbiology and Biotechnology* **19**: 907-912.

Pariza, M.W. and Hargraves, W.A. 1985. A beef-derived mutagenesis modulator inhibits initiation of mouse epidermal tumors by 7,12-dimethylbenz[a]anthracene. *Carcinogenesis* **6**: 591-593.

Pariza, M.W., Park, Y. and Cook, M.E. 2001. The biologically-active isomers of conjugated linoleic acid. *Progress in Lipid Research* **40**: 283-298.

Pariza, M.W., Park, Y., Xu, X., Ntambi, J. and Kang, K. 2003. Speculation on the mechanisms of action of conjugated linoleic acid. In: Sebedio, J.-L., Christie, W. W. and Adlof, R., editors. *Advances in conjugated linoleic acid research*, volume 2: 251-258. AOCS Press, Champaign, Illinois, USA.

Park, Y., Storkson, J.M., Ntambi, J.M., Cook, M.E., Sih, C.J and Pariza, M. W. 2000. Inhibition of hepatic stearoyl-CoA desaturase activity by *trans*-10, *cis*-12 conjugated linoleic acid and its derivatives. *Biochimica et Biophysica Acta* **1486**: 285-292.

Rego, O.A., Rosa, H.J.D., Portugal, P. *et al.* 2005. Influence of dietary fish oil on conjugated linoleic acid, omega 3 and other fatty acids in milk fat from grazing dairy cows. *Livestock Production Science* **95**: 27-33.

Ritzenthaler, K.L., McGuire, M.K., Falen, R., Shultz, T.D., Dasgupta, N. and McGuire, M.A. 2001. Estimation of conjugated linoleic acid intake by written dietary assessment methodologies underestimates actual intake evaluated by food duplicate methodology. *Journal of Nutrition* **131**: 1548-1554.

Rosberg-Cody, E., Ross, R.P., Hussey, S. *et al.* 2004. Mining the microbiota of the neonatal gastrointestinal tract for CLA-producing bifidobacteria. *Applied Environmental Microbiology* **70**: 4635-4641.

Rosberg-Cody, E., Johnson, M.C., Fitzgerald, G.F., Ross, P.R. and Stanton, C. 2007. Heterologous expression of linoleic acid isomerase from *Propionibacterium acnes* and anti-proliferative activity of recombinant *trans*-10, *cis*-12 conjugated linoleic acid. *Microbiology* **153**(8): 2483-2490.

Rowland, I.R., Rumney, C.J., Coutts, J.T. and Lievense, L.C. 1998. Effect of *Bifidobacterium longum* and inulin on gut bacterial metabolism and carcinogen-induced aberrant crypt foci in rats. *Carcinogenesis* **19**(2): 281-285.

Siems, W.G., Grune, T., Hasselwander, O. and Kramer, K. 2001. Conjugated linoleic acid. In: Kramer, K., Hoppe, P.P. and Packer, L., editors. *Nutraceuticals in health and disease prevention*: 257-287. Marcel Dekker, New York, USA.

Smedman, A. and Vessby, B. 2001. Conjugated linoleic acid supplementation in humans – metabolic effects. *Lipids* **36**: 773-781.

Stanton, C., Lawless, F. and Kjellmer, G. *et al.* 1997. Dietary influences on bovine milk *cis*-9, *trans*-11-conjugated linoleic acid content. *Journal of Food Science* **62**: 1083-1086.

Stanton, C., Murphy, J., McGrath, E. and Devery, R. 2003. Animal feeding strategies for conjugated linoleic acid enrichment of milk. In: Sebedio, J.L., Christie, W.W. and Adlof, R.O., editors. *Advances in conjugated linoleic acid research*, volume 2: 123-145. AOCS Press, Champaign, Illinois, USA.

Storkson, J.M., Park, Y., Cook, M.E. and Pariza, M.W. 2005. Effects of *trans*-10, *cis*-12 conjugated linoleic acid (CLA) and cognates on apolipoprotein B secretion in HepG2 cells. *Nutrition Research* **25**: 387-399.

Terpstra, A.H.M. 2004. Effect of conjugated linoleic acid on body composition and plasma lipids in humans: an overview of the literature. *American Journal of Clinical Nutrition* **79**: 352-361.

Thom, E., Wadstein, J. and Gudmundsen, O. 2001. Conjugated linoleic acid reduces body fat in healthy exercising humans. *Journal of International Medical Research* **29**: 392-396.

Topping, D.L. and Clifton, P.M. 2001. Short chain fatty acids and human colonic function: roles of resistant starch and non-starch polysaccharides. *Physiological Reviews* **81**: 1031-1064.

Yang, M. and Cook, M.E. 2003. Dietary conjugated linoleic acid decreased cachexia, macrophage tumor necrosis factor-alpha production, and modifies splenocyte cytokine production. *Experimental Biology and Medicine* **228**: 51-58.

Valeille, K., Gripois, D., Blouquit, M.F. *et al.* 2004. Lipid atherogenic risk markers can be more favourably influenced by the *cis*-9, *trans*-11-octadecadienoate isomer than a conjugated linoleic acid mixture or fish oil in hamsters. *British Journal of Nutrition* **91**: 191-199.

Wahle, K.W.J., Heys, S.D. and Rotondo, D. 2004. Conjugated linoleic acids: are they beneficial or detrimental to health? *Progress in Lipid Research* **43**: 553-587.

Wang, Y. and Jones, P.J. 2004. Dietary conjugated linoleic acid and body composition. *American Journal of Clinical Nutrition* **79**: 1153S-1158S.

Whigham, L.D., Cook, E.B., Stahl, J.L. *et al.* 2001. CLA reduces antigen-induced histamine and PGE(2) release from sensitized guinea pig tracheae. *American Journal of Physiology* **280**: R908-R912.

Whigham, L.D., Higbee, A., Bjorling, D.E., Park, Y.H., Pariza, M.W. and Cook, M.E. 2002. Decreased antigen-induced eicosanoid release in conjugated linoleic acid-fed guinea pigs. *American Journal of Physiology* **282**: R1104-R1112.

Wilson, T.A., Nicolosi, R.J., Chrysam, M. and Kritchevsky, D. 2000. Conjugated linoleic acid reduces early aortic atherosclerosis greater than linoleic acid in hypercholesterolemic hamsters. *Nutrition Research* **20**: 1795-1805.

Council for Responsible Nutrition (CRN) Scientific Event

Nutrition, Healthy Ageing and Public Policy

Royal College of Physician's, Regent's Park, London.

November 2007

Chairman: Professor Peter J. Aggett

Lancashire School of Health and Postgraduate Medicine, University of Central Lancashire, Preston, UK

Abstracts from the Event

The challenges and implications for health policies and nutritional support

David Richardson

DPR Nutrition Limited, 34 Grimwade Avenue, Croydon, UK

Our health is our most precious asset, and the new International Alliance of Dietary/Food Supplement Associations (IADSA) report on nutrition, healthy ageing and public policy is a wake-up call for policymakers and health professionals. Research from around the world has highlighted the nutritional vulnerability of older people and the major lasting impact that poor nutritional status has on their health, well-being and life expectancy. The IADSA report demonstrates that nutrition has the power and promise to mitigate the impending burden of chronic disease and disability as well as to improve the quality of life of this rapidly growing segment of the world's population. Current trends indicate a double burden of under-nutrition on the one hand and obesity, high blood pressure, type 2 diabetes, cardiovascular disease and certain cancers on the other. In addition, changes in body composition during the ageing process (particularly the gradual loss in muscle mass), cataract and age-related macular degeneration (ARMD), the widespread eye diseases associated with diabetes and advancing age, digestive, cognitive and respiratory problems, as well as arthritis and osteoporosis, cause suffering and loss of mobility and independence. Hence, the race is on to determine the role of appropriate nutrition interventions, including food supplements, that may help reduce the risk of these degenerative processes, or at least delay their onset, and may provide the key to reducing healthcare costs. The new report includes several examples where food supplements could make a difference, such as the health benefits of calcium and vitamin D for bone health and reduced risk of osteoporosis, and omega-3 fish oils for cardiovascular benefits. Similarly, multivitamins and minerals could be a prudent choice for people, young and old, to fill the gap between micronutrient intake and the recommended daily amounts. Healthcare and nutrition strategies that modulate favourably age-related declines in most organ functions and reduce the progression of chronic diseases deserve much greater priority in research, information dissemination and education.

The obesity/diabetes/metabolic syndrome challenge: implications for the elderly in Britain

Anthony Leeds

School of Biomedical and Molecular Sciences, University of Surrey, Guildford, UK; Cambridge Health and Weight Plan, Corby, UK

Rising prevalence of obesity and the demographic shift towards an older population indicate that the need for medical care and reduction of risk factors in the over 65 year olds does need to be considered as a special case. Type 2 diabetes is increasing in prevalence but vascular risk factors are aggressively treated with drugs – more people will live longer with diabetes, escaping heart attack but succumbing to renal failure and eye disease. Metabolic syndrome, a combination of risk factors linked to underlying insulin resistance has about a 1 in 5 prevalence among UK adults, is related to body composition and central obesity and can be ameliorated by weight reduction. The combination of the underlying metabolic problem with the inevitable effect of reduced insulin secretion with age presents a challenge: how can diabetes and vascular disease be delayed until late in the now-extended life-span? A re-definition of optimal body composition, risk-factor levels (including glycated proteins) and exercise levels is required. Should the elderly aim for specific levels of abdominal fat and muscle mass? How can this be achieved? What exercise levels are needed and are possible? Can the life-style bad guys' (long working hours, the urban environment, physical inactivity) be brought under control or will the answer be found in not-yet invented medications? Advancing technological development facilitating self monitoring (of several aspects of health) will help some people but not all.

Health-related benefits of n-3 fatty acids (heart, cognitive performance, joints)

Anne Minihane

Hugh Sinclair Unit of Human Nutrition, School of Chemistry, Food Biosciences and Pharmacy, University of Reading, Reading, UK

The amount and composition of dietary fat is recognised as being an important environmental determinant of the development of many chronic disease such as cardiovascular disease, cognitive decline, and autoimmune diseases such as arthritis and asthma. Although the data is not fully consistent, evidence is suggestive that increased intakes of the long chain n-3 PUFA, eicosapentaenoic acid (EPA) and docosahexaenoic acid (DHA), found in oily fish, may reduce the risk and delay the progression of these age-related disorders. In my presentation consideration will be given to: the doses of EPA+DHA associated with particular health benefits; the relative biopotencies of EPA versus DHA; the sustainability of current worldwide fish oil stocks and the potential of alternative sources such as algal or plant sources; current UK recommendations of a minimum of 0.45g EPA+DHA per day; potential toxicological concerns associated with increased oily fish intake; current UK intakes and the use of fish oil supplements and fortification of foods with EPA+DHA as a means to increase population long chain n-3 PUFA intake; and the current evidence regarding the potential use of the shorter chain n-3 PUFA, ?-linolenic acid, as an alternative means of increasing EPA status.

Health benefits of vitamin A and the carotenoids

Hans Konrad Biesalski

Department of Biological Chemistry and Nutrition, University of Hohenheim, Stuttgart, Germany

Vitamin A is essential for growth and development of nearly all cells and tissues. In particular, mucous membranes need this vitamin for adequate function (lung, intestine). The major source is liver, meat and meat-derived food. The mean intake in western countries is 0.8-1.0 mg/day. The recommendation is 1.0 mg/day. Consequently, the supply of natural vitamin A to meet the recommendation is critical, especially in elderly and pregnant women. In a recent study in Germany, we documented vitamin A plasma levels in a deficiency range in 10 out of 36 pregnant women with good social status. A low intake, however, contributes to increased sensitivity against infectious diseases and favours infections of the upper respiratory tract. In addition, lung development of the newborn is critically affected. The supply of provitamin A carotenoids cannot compensate for a low vitamin A intake due to a poor conversion rate. Non-provitamin A carotenoids are also of great importance for human health such as eye health, skin protection and organ function. In contrast to vitamin A, the knowledge of their importance is more or less rudimentary and has to be further elucidated.

Selenium and healthy ageing

Margaret Rayman

School of Biomedical and Molecular Sciences, University of Surrey, Guildford, UK

The essential trace mineral, selenium, appears to be protective in the ageing process. Results from the EVA study in an elderly French population have shown that mortality rates were significantly higher in individuals with low selenium status; when the underlying causes of death were considered, a significant association with cancer-related mortality was found. Furthermore, selenium, which is known to be required by the brain, appears to protect it from the ageing process. Low plasma selenium in the elderly was found to be significantly associated with senility and accelerated cognitive decline, while brain selenium concentration in Alzheimer's patients was only 60% of that in controls. Ageing is associated with a decline in immune response but selenium supplementation appears to be able to restore the age-related decline in immune function to the level found in healthy young people. Cancer is, by and large, a disease of older people. There is growing evidence that higher selenium intakes are associated with reduced cancer risk. Numerous prospective studies have shown that a reduced risk of cancer, particularly of prostate cancer, is associated with a higher selenium intake or status. The Nutritional Prevention of Cancer trial showed that US subjects with a history of non-melanoma skin cancer who were randomised to a daily supplement of selenium (200mcg) had a significantly lower total cancer incidence (37%) and mortality (50%) in a double-blind intervention study when compared to those given placebo. In the context of these anti-ageing effects, low or diminishing selenium status in some parts of the world, notably in some European countries including the UK, is giving cause for concern.

Nutritional aspects of bone health and osteoporosis

Susan Lanham-New

Division of Nutritional Sciences, Faculty of Health and Medical Sciences, University of Surrey, Guildford, UK

Predisposition to poor bone health will result in an increased risk of fracture and hence the clinical diagnosis of osteoporosis. Osteoporotic fracture is a major public health problem, with 1 in 3 women and 1 in 12 men over the age of 55 years affected by the disease at a cost of £1.7 billion per annum. The WHO predicts there will be a projected rise in osteoporotic fracture from 1.66 million in 1990 to 6.26 million in the year 2050. Hence there is an urgent need for public health strategies to target prevention of poor bone health on a population-wide basis. The role that diet plays in the development and maintenance of bone health and prevention of fracture throughout the life cycle remains to be fully quantified. As an exogenous factor, which is (i) amenable to change and (ii) has important practical/public health implications, it clearly deserves special attention. There are good data to suggest that Ca supplements are effective in reducing bone loss in late menopausal women (>5 years post-menopause), particularly in those with low habitual Ca intake (< 400mg/d). The findings of Ca supplementation studies in the early stages of the menopause are conflicting and are an area for further investigation. Vitamin D and Ca supplementation studies have been shown to significantly reduce fracture rates in both institutionalised and free living elderly populations'. Interestingly, vitamin D supplementation alone is not effective. Whilst it is well documented that vitamin D synthesis from sunlight is affected by the ageing process, there is a remarkable lack of awareness of this public health nutrition message. Findings of the National Diet and Nutrition Survey and the 1958 British Cohort show that a large percentage of the population are vitamin D insufficient'. Recent population-based studies suggest a positive association between high intakes of fruit and vegetables and bone health but further data are needed with regard to the mechanisms of action. There are also good data to suggest that vitamin K is critical to bone. The role of other' micronutrients (including magnesium, phosphorus, sodium, fluorine, trace elements, vitamin C, and vitamin B) and other' dietary components (including alcohol, caffeine and phytoestrogens) remains largely undefined and is an area for further research.

A public policy perspective: aligning healthcare and nutrition support

Judith Buttriss

British Nutrition Foundation, 52-54 High Holborn, London, UK

This presentation outlined the approach taken by the British Nutrition Foundation's Task Force on Nutrition and Healthy Ageing, chaired by Professor John Mathers, which was published early in 2008.

Other Speakers

Dis-moi ce que tu manges, je te dirai ce que tu es

Paul Cann

Help the Aged, 207-221 Pentonville Road, London, UK

Panel discussion and Chairman's closing remarks

Printed in the United States
123052LV00007B/2/P

9 780860 141754